RECENT DEVELOPMENTS IN SEWAGE SLUDGE PROCESSING

Proceedings of a workshop held at Lappeenranta, Finland, on 18 June 1990 and jointly organised by

• the Commission of the European Communities, Directorate-General Science, Research and Development

and

• the National Board of Waters and the Environment, Helsinki

RECENT DEVELOPMENTS IN SEWAGE SLUDGE PROCESSING

Edited by

F. COLIN

Institute of Hydrological Research, Nancy, France

P. J. NEWMAN

Water Research Centre, Medmenham, UK

and

Y. J. PUOLANNE

National Board of Waters and the Environment, Helsinki, Finland

ELSEVIER APPLIED SCIENCE
LONDON and NEW YORK

ELSEVIER SCIENCE PUBLISHERS LTD
Crown House, Linton Road, Barking, Essex IG11 8JU, England

Sole Distributor in the USA and Canada
ELSEVIER SCIENCE PUBLISHING CO., INC.
655 Avenue of the Americas, New York, NY 10010, USA

WITH 18 TABLES AND 35 ILLUSTRATIONS

British Library Cataloguing in Publication Data

Recent developments in sewage sludge processing.
I. Colin, F. II. Newman, P. J.
III. Puolanne, Y. J.
628.3

ISBN 1-85166-683-4

Library of Congress CIP data applied for

Publication arrangements by Commission of the European Communities, Directorate-General Telecommunications, Information Industries and Innovation, Scientific and Technical Communication Unit, Luxembourg

EUR 13650

Printed in Northern Ireland by The Universities Press (Belfast) Ltd.

FOREWORD

This volume contains an account of the proceedings of a one day Workshop held at Lappeenranta, Finland on 18 June 1990. The Workshop was held under the auspices of DGXII, Science, Research and Development, of the Commission of the European Communities as part of the Concerted Action COST 681 activity on "Treatment and Use of Organic Sludges and Liquid Agricultural Wastes" (1987-1990).

Under the COST programme five Working Parties have been established to coordinate national research on various aspects of treatment and disposal of sewage sludges and liquid agricultural wastes. Of these, Working Party 1 is concerned with research on the processing of sludges and organic wastes. The present Workshop was organised on behalf of the Working Party by Dr Y J Puolanne from the National Board of Waters and the Environment, Helsinki, Finland assisted by Dr F Colin, Institute for Hydrological Research, Nancy, France and Dr P J Newman, WRc, Medmenham, UK. The overall theme for the Workshop was "Recent Developments in Sewage Sludge Processing" and it considered new developments in two principal aspects of sludge technology - Physical Properties and Disposal Studies.

The first session, which dealt with studies of the Physical Properties of Sewage Sludge, consisted of four papers covering, respectively, assessment of the pollution potential of dumped sludge, defining a standard for the concept of congealed (consolidated) sludge, optimising sludge dewatering using a PC - based operator system and the influence of digestion on sludge characteristics.

The second session, dealing with Disposal Studies of Sewage Sludge, also consisted of four papers, and considered, respectively, economic and quality aspects of composting sewage sludge, co-landfilling and co-digestion of sewage sludge and urban solid refuse, integrated treatment of solid wastes, sewage sludges and liquid manure and integrated solid waste and sludge treatment by the WABIO process.

The timing of the Workshop proved to be very appropriate to help meet the ever-pressing needs of the Water Industry throughout Europe faced with the task of the treatment and disposal of sewage sludge and liquid agricultural wastes. Here the trends are for more sludge to be produced each year and for the restrictions on its disposal, notably in agriculture and to the sea, to become stricter. This will force the Industry to consider new (or at least to reconsider existing techniques in the case of incineration) for the disposal of sewage sludge and liquid agricultural wastes and to greatly improve existing techniques. A better understanding of sludge characteristics will facilitate these developments.

Special thanks are due to Dr Y J Puolanne of the Finnish National Board of Waters and the Environment, Helsinki for his excellent organisation of the local arrangements for the Workshop and also to all the authors and participants for their contributions.

F COLIN

Coordinator, COST 681 Working Party 1

CONTENTS

Session I

PHYSICAL PROPERTIES

Chairman : F. COLIN

Rapporteur : P.J. NEWMAN

2

EXPERIENCES IN THE LABORATORY AND WITH COLUMNS AND LYSIMETERS FOR THE STUDY OF SLUDGE LANDFILLING

F. COLIN

Institut de Recherches Hydrologiques
(I.R.H.)

1. INTRODUCTION

We will here limit ourselves to the study of the following problem: assessment of the pollution potential of dumped sludge. Because of interactions between the solid (very finely dispersed) phase of sludge and interstitial water, sludge may produce soluble pollution, which shall be transported into ground–water.

Our contribution shall consist to establish the synthesis of methodologies, results and observations collected during experiences conducted on various scales: in the laboratory, in small dimensioned columns (a few tens cm in diameter), or in more important (10 cu.m) tight boxes or in lysimeters resting on a pervious soil to which one may have access at various depth levels. These works, – a large part of which is not yet published – were conducted during the last few years.

2. LABORATORY APPROACH

The laboratory is the experimenting level which is best adapted to the fine study of the mechanisms of sludge behaviour and soluble pollution production.

2.1 Analytical characterization

The global analytical characterization of a sludge is indeed necessary as a reference element, but it is totally insufficient for foreseeing the immediate or long–dated behaviour of the sludge.

The separate analysis of the constituent phases (solid and liquid phases), after their separation and the measure of their relative proportions, furnishes more information by considering a mobility factor. The liquid phase my indeed be considered as mobile, and its contents can be directly washed away by infiltration water. On the contrary, the pollution contained in the solid phase is little mobile and shall be only more slowly liberated by physical solubilization, hydrolysis, biodegradation, etc...phenomena.

2.2 Characterization of sludge leachability

The immediately soluble fraction of the solid phase of sludge can be appreciated by means of leaching tests, consisting in one or several operations putting this solid phase into contact with an aqueous solution (lixiviant) and gathering lixiviates which will be analyzed. It is desirable to effect several successive operations, in order to deplete the solid phase or to reach a constant degree of solubilization.

A large synthesis effected in 1985 (1) describes in detail the problematics of leaching tests and the numerous already applied methods. These methods are arbitrarily defined and are greatly different in their parameters and operating factors :

- nature of leaching solution,
- solid/liquid ratio,
- contact duration,
- number of successive contacts (elutions),
- mode and intensity of agitation.

Techniques applicable to waste materials of any nature are standardized or approved in numerous countries (for instance Ref. 2) Without being exhaustive, document (1) lists 30 different procedures.

A mass balance of the various conducted operations enables to calculate the total solubilized quantities relative to the studied sludge mass. For instance, Table No. 1 gives such a balance realized for an anaerobic digested urban sludge, which was mechanically dewatered after coagulation by ferric salts (dryness = 36.3 %; volatile solids = 35.3 % of dry solids; leaching ratio S/L = 1/10; lixiviant = demineralized water).

Elements	Extracted quantity by the first extraction (mg/l)	Extracted quantity by the 2nd extraction (mg/l)	Extracted quantity by the 3rd extraction (mg/l)	Total extracted quantity per kg raw waste material (mg/kg)
Cadmium	< 0.01	< 0.01	< 0.01	< 0.1
Total chromium	< 0.02	0.06	0.03	0.9
Copper	0.05	0.03	0.06	1.4
Iron	3.75	13.5	6.75	240
Manganese	1.12	2	1.06	41.8
Nickel	< 0.05	< 0.05	< 0.05	< 0.5
Lead	0.10	< 0.05	0.10	2
Potassium	9	8	2	190
Sodium	9	2	1	120
Zinc	0.26	0.19	0.31	7.6

Table 1 – Balance of successive elutions by the leaching test of an anaerobic digested sludge

The obtained results depend largely on the nature of the lixiviant. This is either arbitrarily defined, or designed in order to simulate the liquid aqueous phases with which the waste material may be brought into contact in the natural environment or within the landfill.

Table No. 2 compares the leached quantities from an anaerobic digested sludge, dewatered by a pressure belt filter, the following solutions being utilized as lixiviants :

- demineralized water,
- a very dilute synthetic solution simulating contaminated rain water (saturation in O_2, CO_2, traces of HNO_3 and H_2SO_4),
- a solution loaded with products form the aerobic degradation of organic matter (aggressive free CO_2),

- a solution loaded with intermediate metabolites (ammonium acetate and acetic acid) from the anaerobic digestion of organic matter.

Lixiviating solution / Elements	Demineralized water	Simulated rain water	Water loaded with anaerobic degradation products of organic matter (CO_2)	Water loaded with anaerobic degradation products of organic matter (volatile acids and their salts)
Ag	< 0.02	< 0.02	< 0.02	< 0.02
Cr	< 0.02	< 0.02	< 0.02	0.05
Cu	0.03	0.03	0.03	0.03
Fe	0.4	0.35	1	18
Mn	0.03	0.04	0.17	0.61
Ni	< 0.05	0.08	< 0.05	0.20
Pb	< 0.05	< 0.05	< 0.05	< 0.05
Zn	< 0.02	< 0.10	< 0.02	0.26

Table 2 – Comparison of the concentration of the first leachates (in mg/1 of an anaerobic digested sludge, in function of the nature of lixiviant

Leaching tests enable at the best to forecast the immediate or very short–dated liberation of pollution, as they de not take into account the phenomena of longer–dated evolution.

2.3 Study of the long–time behaviour

In the case of sludge from urban origin and comparable sludges, this evolution is essentially of a biological nature.

The biodegradation of organic matter can liberate elements (metals and non–metals) which it had fixed. On the other hand, it may act on the solubilization of the mineral phase, as it can deeply modify the environmental conditions in terms of acido–basicity and oxidation–reduction.

For solid waste materials, we have developed techniques which apply respirometric devices enabling, in aerobic medium, to measure the oxygen consumption with time of the seeded waste, or the quantity of carbon dioxide produced through the mineralization of organic matter; or, in anaerobic medium, to determine the volume and the composition of the gaseous phase (mixture of methane and carbon dioxide produced by the biodegradation). Fig. 1 shows the scheme of a device for the study of the aerobic behaviour of a sludge : an air flow, which previously has been deprived of CO_2 and moistened, crosses the sludge and takes away the CO_2 produced by biodegradation; this CO_2 is then captured by a washer bottle containing a soda solution, which is periodically titrated and renewed. Fig. 2 shows an example of mineralization curves obtained on various types of urban and industrial sludges.

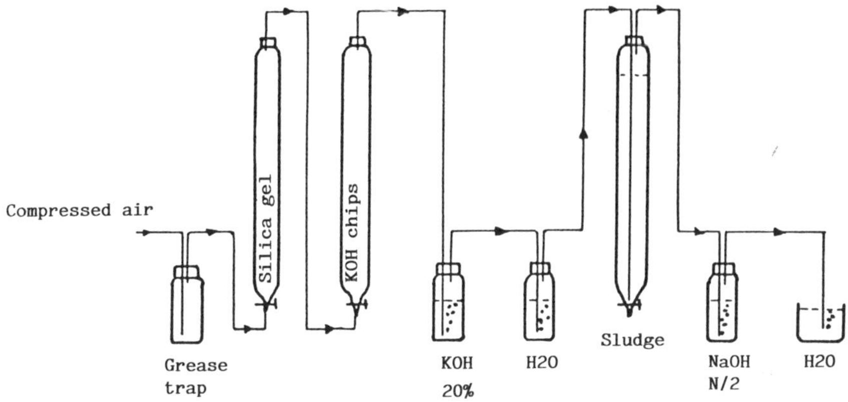

Fig. 1 – Respirometric device for studying the aerobic degradation of a sludge

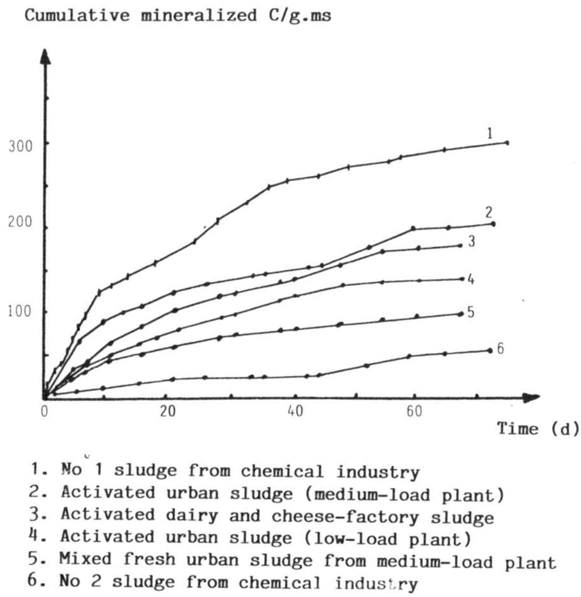

Fig. 2 – Mineralization curves for various kinds of sludge

The global respirometric observation of sludge mineralization may be supplemented by the measurement of various biochemical parameters which characterize its biological stability : measure of the present biomass through ATP determination, measure of the tendency to solubilization of the solid organic matter through enzymatic hydrolysis, characterized by the measure of some typical hydrolytic enzymatic activities. For instance : proteolytic, amylasic, cellobiasic activities, such as illustrated in Figs. 3 and 4 by the aerobic and anaerobic evolution of a fresh sludge.

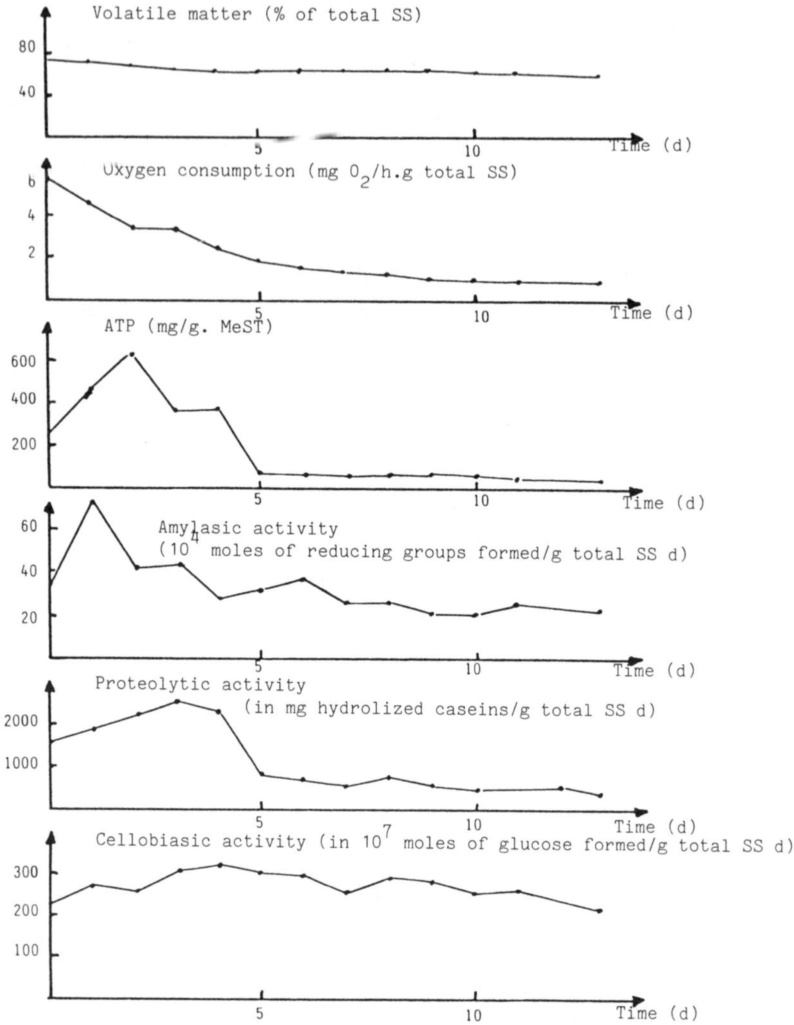

Fig. 3 – Trends in the descriptive parameters of the biological stability of sludge during aerobic behaviour at 20°C

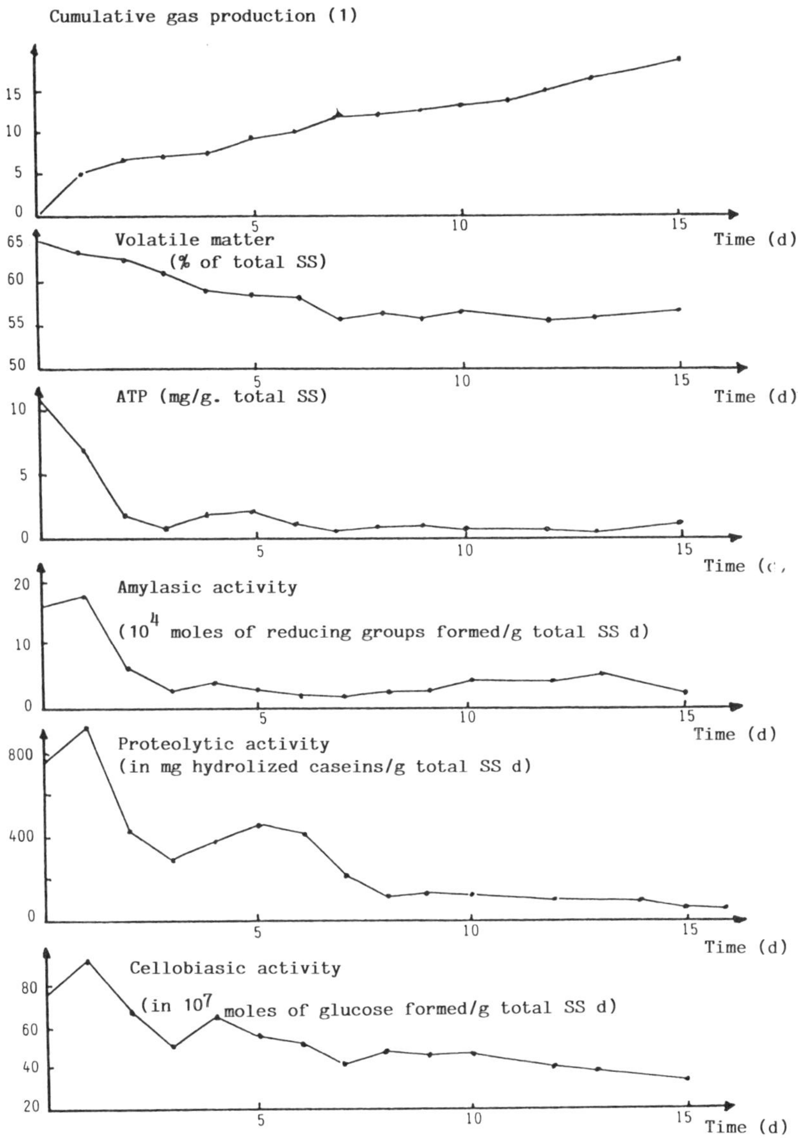

Fig. 4 – Trends in the descriptive parameters of sludge stability during anaerobic behaviour at 35°C

When test parameters (temperature, presence of nutrients, redox conditions, effects of various limiting factors) are optimized, these tests enable to hasten the evolution of sludge and to rapidly aim to its final state in aerobic or anaerobic conditions. It is then possible to conduct an analysis of the liquid phase of the final sludge and to effect leaching tests on the residual solid phase in order to determine the quantities of polluting elements which were solubilized or made able to be solubilized during the total biological transformation of this sludge.

In spite of their accelerated character, these tests must have a duration from several weeks to a few months according to the nature of the material.

2.4 As a conclusion, according to the arbitrary character of their design or their applied conditions, laboratory tests are able chiefly to assess a pollution potential of the landfilled sludges, but are unable to forecast the liberation kinetics of this pollution. This objective requires experimental conditions nearer to those to which the landfilled sludges are effectively exposed.

3. EXPERIMENTING IN PILOT COLUMNS

These experiences are destined to satisfy, at least in part, the above objective. Their principle consists to place the sludge (and the material with which the sludge generally must be dumped for reasons of mechanical stability) in a column which, according to the case, shall be directly exposed to the weather conditions (temperature, pluviometry) or, on the contrary, submitted to controlled conditions of water transfer through spraying and keeping at a constant temperature.

The first of these options corresponds to more realistic conditions, but does not allow the reproducibility of tests at different periods. This risk may be minimized by using always identical control columns which, in this case, will somewhat facilitate the interpretation of results. This type of test shall then be effected with practical objectives of evaluating the behaviour of sludges or waste materials, chiefly in view of the decision of admissibility of certain industrial sludges in dumping sites.

On the contrary, the tests in well controlled conditions will be reserved for objectives of fine study of the evolutive mechanisms of landfilled sludges, in view notably of modelizing.

Numerous works were conducted for studying the co–landfilling of sludges and household refuse. We will describe here an example of experimenting and its results. The effected study (3) related to 13 waste sludges belonging to 4 classes of waste materials from urban sewerage :

- sludge from the cleaning of combined or separate sewerage systems,
- waste sludge from municipal waste water treatment plants,
- sludge from grease traps of urban or community treatment plants,
- fecal substances from individual sewerage.

The tests were effected in columns with a diameter of 40 cm, exposed to atmospheric conditions. Sludges were introduced at the top of a 90 cm layer of old crushed household refuse. This refuse was already largely stabilized, as it itself evoluted for 3 years in 10 cu.m boxes exposed to local weather conditions. Experimenting lasted 600 days and consisted to observe the production of percolate with time for every column and for a control column which received only the supporting material (crushed household refuse). Percolates were analysed every week, and the cumulated quantities on stripped pollutants in function of time were represented. A preliminary analysis on the initial sludges and household refuse, as well as leaching tests, enabled to establish a mass balance of the fate of sludges during the 600 days of experimenting.

Fig. No 5 and Table No 3 give detail examples of the obtained results.

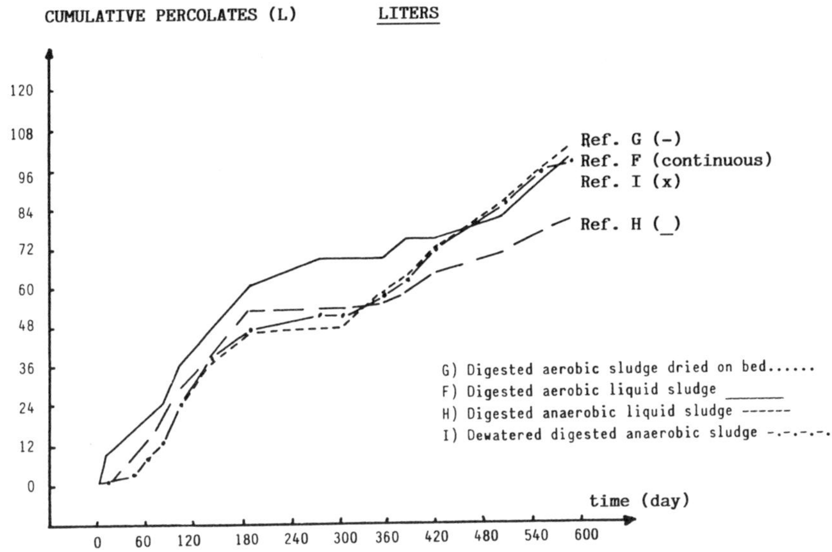

Fig. 5 – Residual urban sludge. Cumulative quantities of percolates

Reference of column / Determination	Control (household refuse)	Urban waste sludge			
		Réf. F	Réf. G	Réf. H	Réf. I
Percolates (litres)	87	99	101	80	96
COD (g 02)	12	12	22	9.7	27
BOD5 (g 02)	0.59	0.51	0.61	0.36	1.59
Total NH4 (g NH4+)		0.36	0.31		
Kjeldahl-N (g N)	0.91	3.03	1.47	2.0	2.60
Dry matter (g)	152	172	494	184	471
Cadmium (mg)	0.5	0.6	0.8	1.9	0.1
Total chromium (mg)	1.8	2.1	1.4	2.3	1.3
Copper (mg)	3.8	2.0	6.2	3.4	6.8
Iron (mg)	34	12	12	17	11
Manganese (mg)	27	18	32	55	32
Nickel (mg)	5.6	9.3	8.9	9.5	13
Lead (mg)	7.3	2.4	12	9.1	13
Zinc (mg)	2.9	6.5	2.7	3.1	4.7
Cobalt (mg)	2.4	1.3	2.0	2.7	3.5

Table 3 – Global quantities of matter solubilized during 600 days experimenting

More globally, the study led to following conclusions :

– In a certain number of cases and for certain pollutants, the quantities of solubilized pollutants are lower than by the dumping of the supporting material alone. This is due to interactions of the waste material from urban sewerage introduced with domestic refuse : action on the volume of percolates, on the retention and migration characteristic of water, adsorption phenomena (grease), eventual chemical and biological reactions. This observations shows that it is difficult to set a mass balance of the fate of polluting elements brought by the refuse, as at the same time the quantity of pollution liberated by the supporting material could vary.

– In the case of sludge from sewer cleaning, the dumping in common with domestic refuse is accompanied by an important increase of the organic pollution (COD and, at a lesser degree, BOD_5) liberated for :

. grit chamber sediment from a combined sewerage,

. sludge from the cleaning of a combined sewerage,

. sludge from gullies of a storm–water drainage system.

On the contrary, pollution is lower than for the control sample in the case of sludge from gullies of a storm–water drainage system.

The liberated quantities of metallic elements are in all cases very low, and of the same order as for domestic refuse alone.

– In the case of urban waste sludges, solubilization of BOD_5 and COD remain of the same order as in the case of domestic refuse alone. They are more important in the case of sludge in solid state (drying beds or mechanical dewatering), but this is due only to the use of quantities of organic matter more important than in the case of liquid sludge. The solubilization of pollution, referred to the mass of dry sludge, is lower for solid sludge than for liquid sludge.

Metal solubilized quantities are very low and of the same order than for the control : more important for certain metals and sludges, and less important for other ones. No general conclusion can thus be derived therefrom.

– In the case of sludge from the grease–trap of urban treatment plant the solubilized quantities are close or even frequently lower than those for domestic refuse alone. On the contrary, for the sludge from the grease–trap of collective restaurant installation, pollution solubilized quantities are from 15 to 100 times more important :

. fecal substances have the same behaviour as liquid waste sludge from treatment plants,

. the pollution quantities liberated in 600 days are 10 to 100 times lower than quantities which can be solubilized by the leaching test applied at the laboratory scale (French method).

The limits of the experiences in pilot columns lay on the level of representativity from the hydraulic point of view (possibility of preferential flows near the wall and relative importance of such flows; sometimes insufficient ration between column diameter and dimension of chips in the matrix of waste material). For the thermal aspect, inertia is low in comparison with the inside of a landfill site, and does not allow phenomena of thermal evolution to manifest themselves (heating due to fermentation). On the contrary, the importance of experimenting allows precise mass balances and the access to the whole concerned material (realization of tappings for collecting samples).

4. EXPERIMENTING IN LARGE PILOT BOXES OR LYSIMETERS

The experimenting scale is definitely larger than in the previous cases (several tens to hundreds cu.m), but at the price of a much more important investment and a larger experimenting clumsiness.

A distinction must be made between tight boxes, allowing to wholly collect percolates, and bottomless boxes which rest on the substratum and require a special outfit for observing the movement of percolates in the subsoil and take samples there.

4.1 Study of the influence of the nature and disposition of an additional material dumped with sludges

We have effected a research (4), the objective of which was to study the influence of the nature and the co–landfilling of an additional material on the pollution solubilized and carried away from dumped sludges. The studied sludge came from a metal surface treatment industry and contained typical pollutants (heavy metals) of well known nature and speciation. Fundamentally different model additional materials were selected :

- an inert and incompressible mineral material : sand of fine grain–size,
- an organic and hydrophillic material, slowly biodegradable and compressible : sawdust.

For a same volume application of both constituents relatively to the sludge mass, various dispositions were studied :

- intimate mixing sludge/additional material, alternate layers of same or different thickness.

The most evident conclusions to be drawn at the end of this research are as follows :

- The use of an intercalated material, for dumping mechanically dewatered sludges containing metal compounds has a significant action on the water balance during the first months of dumping. In this respect, the use of a material of organic nature, with a large water holding capacity and compressible, enables to cut down the flow peaks of infiltration water and to fix a quantity of water corresponding to several months rainfall. This effect is attenuated in the case of the use of a mineral material with a low water holding capacity, of which sand is a typical example. The disposition, as a homogenous mix, of sludge and intercalated material is less efficient than the stratified disposition for minimizing the volume of infiltration water during the first months of dumping. In the case of stratified disposition, it is desirable to realize sufficiently thick layers of intercalated material (as thick as sludge layers for instance).
- At the end of a period of 1 to 2 years, the intercalated material reaches a relatively stable moisture and does no more affect the production of infiltration water; the main interest of using an appropriately selected intercalated material thus seems to spread in time the progress of the pollution peak corresponding to the stripping of the fraction of the waste material which is immediately soluble in water.
- The various metallic elements have a different behaviour: in the case studied here, iron and manganese are more solubilized when an organic intercalated material is used rather a mineral one, and a reverse tendency appears for nickel an chromium because of a specific fixing phenomenon by the utilized organic material (sawdust).

The use of an intercalated material of organic biodegradable nature confers to infiltration water a more acidic character, with shifts which may amount up to 1.5 pH units during the first months. This difference then decreases, without disappearing, during a long period (higher than the duration of our experimentation, i.e. 25 months), with a rest deviation of about 0.5 pH unit. This phenomenon may exert only an incidence on the solubilization of metallic elements, when these are not fixed by the intercalated material or when the latter reaches saturation.

- Pluviometry acts not only on the flow of infiltration waters, but also on their concentration; concentration peaks being observed after dry periods. One thus may hope to minimize pollution carrying away by optimizing the disposition and the selection of intercalated materials able to regulate the flow of infiltration water. Supplementary works in this field should be desirable relating to an evaluation of instantaneous rather than average flows and concentrations, in view to precisely determine the variation with time of the flux : (flow x concentration) product, which corresponds to the carried away quantity of metallic elements.
- In the case of sludge containing metal derivates, and produced by the neutralization of acidic effluents with a base, the tendency of certain metals to be solubilized by lixiviation seems to increase in function of the dumping time, this being the consequence of the stripping of the base initially present in excess, and therefore of the pH decrease which it causes.
- The selection of low cost intercalated materials (waste substances or products without significant value), possessing specific properties for fixing certain metallic elements, can be encouraged by the dumping of refuse containing essentially these elements (for instance : sawdust

in the case of chromium rich waste materials). This process, on the contrary, is open to criticism in the case of refuse containing non–fixable metals, the solubilization of which may increase.

4.2 Study of the common landfilling of industrial sludges with domestic refuse

The objective of this research was to assess the interactions between dumped sludge and additional material, in the case where the latter is constituted by domestic refuse, notably for evaluating the potential of the additional material for the retention of the pollution generated by sludge.

These works were conducted jointly by I.R.H. on the level of experimenting in tight boxes (5) and by B.R.G.M. in full size, with use of refuse of the same nature and origin.

The result of this study were already reported (6), so we will limit ourselves to a short summary of the description of the works and the reached conclusions.

Five sludges from various origins (grease–trap from an urban treatment plant, hydrocarbon wastes, metal surface treatment, formophenolic containing waste water, and sludge from wool degreasing), were incorporated or spreaded on crushed domestic refuse, and then studied for several years, first in boxes (unit volume = about 10 cu.m) for 1365 days, then on experimental plots (unit volume = 450 cu.m) situated on a semi–permeable dumping site. Following the first results, the applied sludge proportions (about 100 kg sludge/m^3 refuse in the boxes) were further reduced for the experiments on the site (between 18 kg/m^3 domestic refuse for urban degreasing sludge).

The obtained results show the abilities of domestic refuse to fix or remove pollutants; for instance : temporary retention (but no real immobilization) of certain forms of pollution (organic matter, hydrocarbons for refuse containing an oily fraction) and permanent retention of metals (notably in sludges from metal surface treatment with an important proportion of chromium, iron or zinc in the precipitated state). Moreover, the important liberation of phenol observed in box could be reduced on the site, by lowering the transfer doses of formophenolic waste water.

Endly, as to the hydrologic balance of the system, the liquid fraction of sludges spreaded in favourable weather conditions (Summer) underground.

5. CONCLUSIONS

The various given examples show the planes of a possible approach for the study and forecasting the liberation of dissolved pollution by landfilled waste sludges.

When these techniques may be applied to ordinary urban waste sludges, they are still much more interesting for this type of sludge when it is abnormally contaminated by pollutants from industries or crafts, and with greater reason for sludges from industrial origin, and chiefly those which contain toxic or dangerous pollutants.

The three levels of approach : laboratory tests, tests in pilot columns, and use of large boxes or lysimeters, bring complementary informations and should logically be applied in the given order, the transition from a stage to the following one being characterized by increasing costs and time of response, but also by an increasing richness and validity of information.

Reference (7) gives the identification of tests conducted on these various scales on different waste materials, as well as the discussion of the comparison of the corresponding results.

Acknowledgements

The author wishes to express his thanks to the French "Secrétariat d'Etat chargé de l'Environnement" SRETIE, whose financial assistance, in the frame of the "Comité Sol–Déchets", enabled the realization of the works the results of which are reported here.

REFERENCES

(1) COLIN F. (1986) "Etude des mécanismes de la génèse des lixiviats. Inventaire et examen critique des tests de laboratoire" in "Etude bibliographique sur les lixiviats produits par la mise en décharge de déchets industriels". CCE XII / 20 / 86 volume 1/4, pp. 1–143.

(2) AFNOR (1988) "Déchets. Essai de lixiviation". Norme expérimentale X31–210.

(3) COLIN F. (1985) "Etude de la mise en décharge des déchets de l'Assainissement Urbain". Compte-rendu d'une recherche effectuée avec l'aide du Ministère de l'Environnement. Comité Sol–Déchets. Contrat n° 237–01–83–40–048. Rapport IRH Nancy n° 85–149.

(4) COLIN F. (1980) "Etude du devenir de l'eau et de son interaction avec les déchets dans une décharge admettant des déchets industriels homogènes (boues d'épuration d'effluents industriels)". Compte-rendu d'une recherche effectuée avec l'aide du Ministère de l'Environnement. Comité Sol–Déchets. Contrat n° 9242. Rapport IRH Nancy n° 80–40.

(5) COLIN F. (1985) "Mise en décharge mixte de boues ou déchets liquides et d'ordures ménagères". Compte-rendu d'une recherche effectuée avec l'aide du Ministère de l'Environnement. Comité Sol–Déchets. Contrat n° 4039. Rapport IRH Nancy n° 85–143.

(6) BARRES et al (1988) "Etudes expérimentales de mise en décharge conjointe d'ordures ménagères et de boues industrielles (décharge mixte)". ISWA 88 5th International Solid Wastes Conference Copenhagen 11/16 Sept. 1988.

(7) NAVARRO A. et al (1986) "Comparaison des résultats d'essais en laboratoire, à l'échelle semi-grand et en vraie grandeur". In "Etude bibliographique sur les lixiviats produits par la mise en décharge de déchets industriels" CCE XII / 20 / 86. Volume 1/4, pp. 145–321.

MECHANICAL STRENGTH DETERMINATION OF COHESIVE SLUDGES - A BELGIAN RESEARCH PROJECT ON SLUDGE CONSISTENCY

P. GEUZENS
SCK - Nuclear Research Centre
W. DIELTJENS
SVW - Centre for Water Research
BELGIUM

Summary

In the Belgian Flemish region a great part of the sludges emerging from sewage treatment and drinking-water production are disposed of at sanitary landfills. One of the requirements that must be met by dumped sludges relates to its consistency.

Companies involved with sewage treatment and drinking-water production have ordered a research project regarding the consistency of their dewatered sludges. This research is aimed at defining - and evaluating the consequences of - a standard for the mechanical strength of sludge, to be disposed of at sanitary landfills. Measurements of the consistency of the sludges are performed by means of the laboratory vane test and a fall cone penetrometer.

In the Belgian Flemish Region sewage treatment is undertaken by one public company, the VMZ. Three companies are active in the production of drinking-water, namely AWW, PIDPA and VMW. Both activities lead to the production of considerable amounts of sludge. A great part of it is dumped at sanitary landfills.

The sanitary landfills in the Flemish Region are distinguished into different classes. Drinking-water sludge and sewage sludge mostly may be dumped on class 2 landfills. These are primarily intended for the disposal of municipal refuse. Sometimes however the contamination of the sludges with heavy metals will make it necessary to dump on class 1 landfills. These are provided for the disposal of industrial, non-toxic wastes.

Flemish legislation requires that waste materials dumped on sanitary landfills are "congealed". The term "congealed" however is imprecisely defined. This gap in the law has already lead to difficult juridicial conflicts.

The governmental organisation OVAM controls the disposal and treatment of waste materials.

VMZ, the Flemish drinking-water producers and OVAM have joined forces in the Centre for Water Research for their common scientific and practically oriented research.

In october 1989 this research centre started a project on the consistency of sludge, with special regards to the drinking-water and the sewage sludges. The project could lean on the results of a preceding literature study on determination methods for geomechanical properties of sludges (1).

1. Aims of the project

A definition and numerical interpretation of the term "congealed" is important for different parties. The producer of sludge may develop a purposive research and management of the sludge dewatering process. Those responsible for the dumping sites will be able to more easily decide on the acceptance or refusal of certain sludges. The controlling authorities may better define and regulate the exploitment conditions.

The aims of the project are :

- to conduct a circumstantial experimental research towards the influence of the origin of the sludge, the dewatering technique and added materials on the characteristics of the sludge, and on its suitability for landfilling;
- to propose a regulation for the concepts of sludge-consistency, possibility to override and stability of the landfill. The consequences of such a regulation should also be evaluated technically, economically and ecologically.

LEGEND

No	= no dewatering	AGR	= agriculture
LAG	= lagoon	B.E.	= black earth
DB	= drying bed	LF 2	= class 2 landfill
FP,L	= chamber filter press lime	LF 1	= class 1 landfill
FP,P.E.	= chamber filter press poly-electrolyte	INC	= incineration
SBP	= sieve belt press		
CEN	= centrifuge		

	AGR	B.E.	LF 2	LF 1	INC	total	in %
No	4547	-	16	-	197	4760	18%
LAG	15	-	-	-	-	15	0%
DB	350	184	1121	157	-	1812	7%
FP,L	161	912	2390	1268	1661	6392	25%
FP,P.E.	-	-	4730	1908	1524	8162	31%
SBP	325	712	374	35	-	1445	6%
CEN	-	-	282	286	2880	3447	13%
Total	5398	1808	8912	3653	6261	26032	100%
in %	21%	7%	34%	14%	24%	100%	

Table 1 : Dewatering and destination of sewage sludge. Amounts in tons of dry matter, 1988.

2. Sludge production in the Flemish Region

An inventory was made of the amounts of sludge that are produced in sewage treatment and in drinking-water production in the Flemish Region.

Sewage sludge

The following dewatering methods for sewage sludge are presently at use :

- sludge ponds or lagoons : large ponds in which the sludge sediments and consolidates under its own weight,
- sludge drying beds : water drains out of the sludge, that is initially spread to a height of about half a meter. At the same time the sludge may dry in the air. Depending on the weathering conditions and the residence time of the sludge high dry matter contents may be reached. The obvious disadvantage of the method is the space it requires.
- mechanical dewatering with :
  - sieve belt presses
  - centrifuges
  - chamber filter presses.

With respect to landfilling the degree of dewatering and the degree of stabilisation are important features of the sludge (consistency, odour emissions). The heavy metal content also determines the choice between class 1 and class 2 landfills.

The total disposal of sewage sludge produced amounted to 26.000 tons of dry solids in 1988. Table 1 and figures 1 and 2 show the subdivision of this quantity according to dewatering method and destination of the sludge. 50 % of the sewage sludge produced was dumped on landfills. More than 50 % of the sludge was dewatered with chamber filter presses.

According to the VMZ the specific sludge production may be estimated at 13,1 kg dm/IE.year.

Drinking-water sludge

Two sources are available for the production of drinking-water : surface waters and ground water.

Ground water in the Flemish region almost always contains too much iron and therefore must be treated. This produces "de-ironing" sludge by flocculation.

Storage basins are always used in the production of drinking-water from surface water. In these basins sedimentation takes place, producing sludge that is cleared out only after longer periods (typically every five years). The purification and treatment of the water produces flocculation sludge.

Drinking-water sludge has a mainly inorganic composition, in contrast to sewage sludge.

Dewatering of the flocculation sludge takes place by lagoons or by mechanical dewatering processes. The mechanical techniques used are the same as with sewage sludge : chamber filter presses for sludge from surface water, centrifuges and belt presses for sludge from ground water.

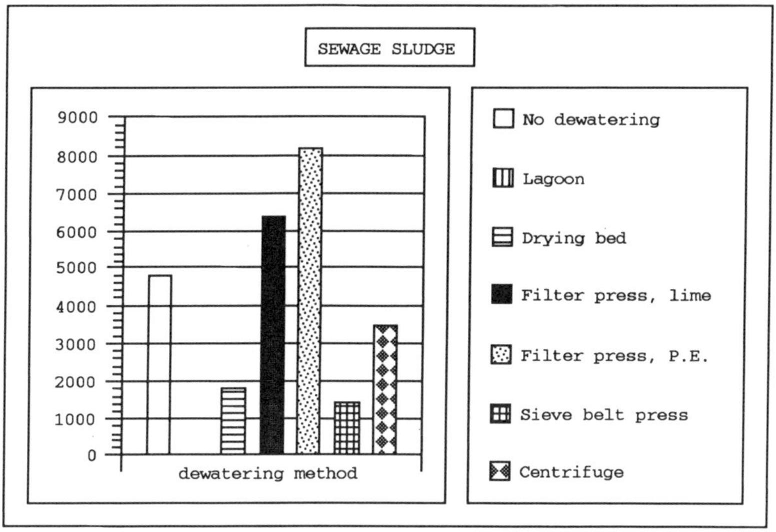

Figure 1 : Dewatering of sewage sludge.

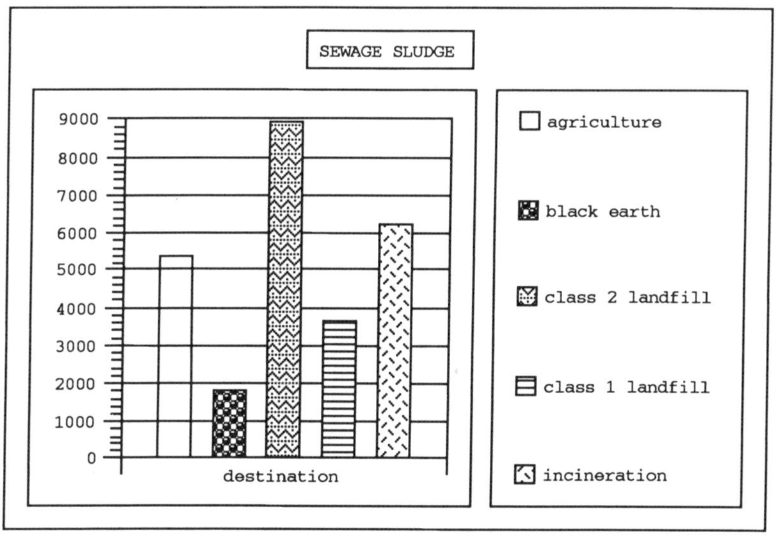

Figure 2 : Destination of sewage sludge.

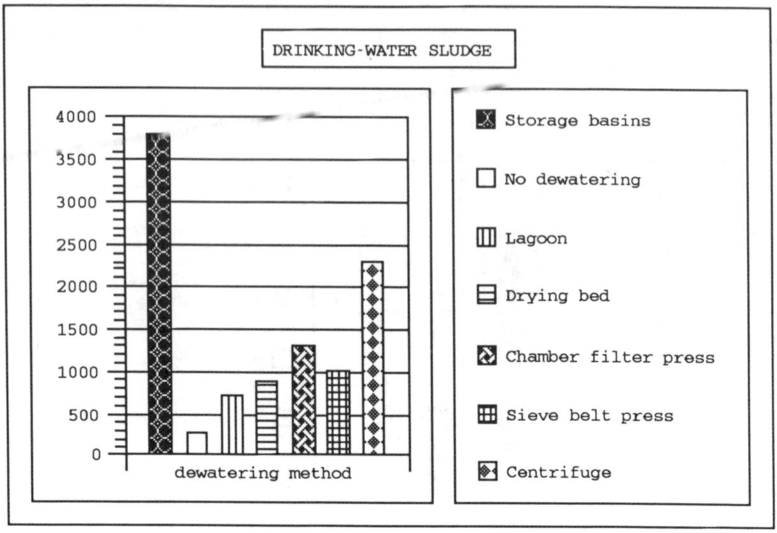

Figure 3 : Dewatering of drinking-water sludge.

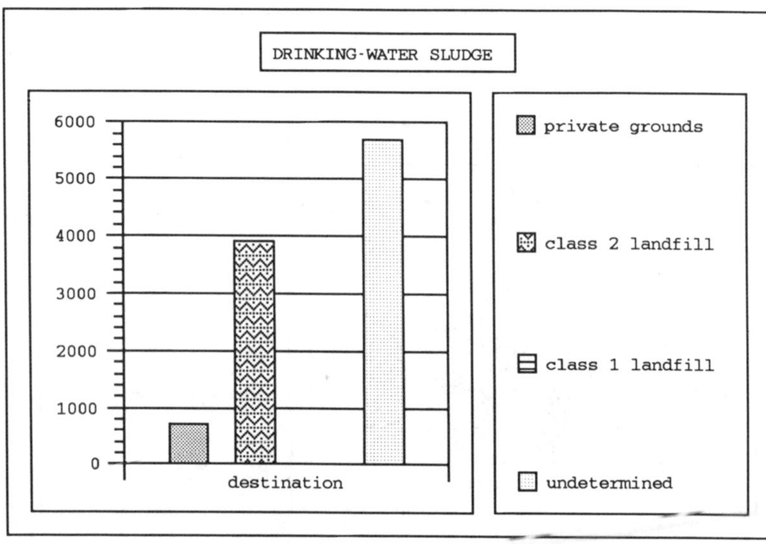

Figure 4 : Destination of drinking-water sludge.

For the disposal of the drinking-water sludge only landfilling is eligible. In the past this used to be done mainly on private grounds of the drinking-water companies (mono-landfills). At present most sludge is disposed of on class 2 landfills. Until now sludge from the storage basins was disposed of in accidental groundworks. This disposal route may in the future also be blocked. The amount of sludge accumulating in the basins is not yet precisely known for all drinking-water plants. The amount given in table 2 and figure 3 must probably be doubled.

LEGEND

BAS	= storage basins	PRIV	= private ground
No	= no dewatering	LF 2	= class 2 landfill
LAG	= lagoon	LF 1	= class 1 landfill
DB	= drying bed	UND	= undetermined
FP	= chamber filter press		
SBP	= sieve belt press		
CEN	= centrifuge		

	PRIV	LF 2	LF 1	UND	total	in %
BAS	-	-	-	3807	3807	37%
No	-	-	-	285	285	3%
LAG	-	-	-	725	725	7%
DB	-	-	-	886	886	9%
FP	700	600	-	-	1300	13%
SBP	-	1008	-	-	1008	10%
CEN	-	2312	-	-	2312	22%
Total	700	3920	0	5703	10323	100%
in %	7%	38%	0%	55%	100%	

Table 2 : Dewatering and destination of drinking-water sludge. In tons of dry matter, 1988.

3. Sludge on sanitary landfills in the Flemish region

At present 13 public class 2 landfills and 4 public class 1 landfills are active in the Flemish region. 4 class 2 landfills do not accept sludge.

Based on literature and preliminary conversations with dumping site responsibles a questionnaire was made which asks for

- general information regarding the dumping site
- dumping techniques and way of treating sludge
- problems on the dumping site with sludge
- criteria used by the dumping site responsibles to accept or reject sludges.

This questionnaire was presented at the exploitants of all sludge-treating and public class 1 and class 2 landfills in the Flemish region.

The most important consequences of this research are :

- The relative amount of sludge supplied at class 2 landfills mostly is less than 5 % (with one exception). On class 1 landfills this relative amount may be more than half the total amount of waste supplied. A great part of this class 1 sludge however is solidified prior to dumping (added materials are fly ash, cement and lime).
- Depending on the relative amount of sludge supplied, the sludge is build in either in thin layers which leads to a thorough mixing with the surrounding waste, or in packages or layers of up to 2 meters high.
- Slipping of waste-supplying trucks on the sludge is most often mentioned as the primary problem with sludge. This is closely related to the pressing out of the sludge towards the surface. Odour emissions are also often mentioned as a problem.
- Class 2 landfills evaluate the consistency of certain, well known sludge streams on the practical problems or possibilities they experience with these streams on the landfill. The need for a generally accepted measuring method for the consistency of sludge is especially felt by class 1 landfill responsibles and undertakings that solidify sludge.

4. Systematic sampling and analysis

38 sludge production places, provided of an installation for mechanical dewatering or of sludge drying beds, were chosen for a systematic and periodical sampling of the dewatered drinking-water or sewage sludge. Recently 7 industrial waste water treatment plants, spread over different industrial activities, were also taken into account.

The following fysico-chemical analyses are performed on these sludges :

- dry matter content
- glow loss
- fermenting material (EAWAG-procedure)
- carbonate content
- sand content (inert materials)
- Fe
- Mg
- Ca
- Al

Two mechanical tests are used to determine the consistency of the sludges :

- laboratory vane test
- Indaver-test (fall cone penetrometer)

On a restricted number of sludges a pocket penetrometer test will be applied as well.

The laboratory vane test has been described repeatedly, also in the COST 681 context (see for example Otto-Witte in (2), p.309). The most recently published proposal for execution of the test is followed (3).

The second mechanical test is a fall cone penetrometer test. This test was developed by the responsible of an industrial landfill in Belgium who perceived the gap in the law and therefore took the initiative to propose a definition of the term "congealed" (4). The test is meant to be applicable to all kinds of waste materials, thus not only to cohesive sludges for which the vane test, in an international context, is more often used.

The test is based on the use of a high precision penetrometer in conjunction with a specifically designed fall cone. A container of heigth ±5 cm and diameter ±5 cm is filled with the sludge which is gently pressed by hand. The calibrated fall cone is placed on top of the equalised surface and allowed to fall for 12 seconds. The penetration depth is then accurately measured. As with the vane test an average of three trials is taken as the result. The dimensions of the fall cone used are given in figure 5. This test is named the Indaver test after the landfill where it was developed.

Along with the test went a criterion for the definition of "congealed" materials which was formulated as follows :

- when the dry solids content is >80% the material always is congealed;
- when the dry solids content is <20% the material never is congealed;
- when the dry solids content is between 20 and 80% the consistency is determined by the results of the penetrometertest : a penetration depth <15mm gives a congealed material, a penetration depth >15mm gives a non-congealed material.

This criterion was developed on the basis of the practical experiences with different sludges on the landfill. These findings however were never published.

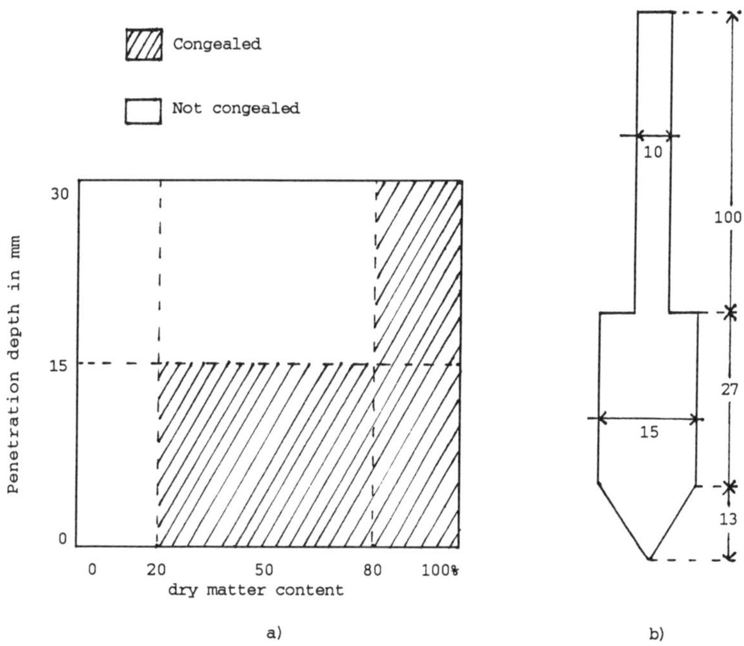

Figure 5 : The Indaver test : a) criterion and b) dimensions of the fall cone in mm. The fall cone is made of stainless steel and weighs 105,6 g.

Figures 6 and 7 give some impression of the results obtained until now. Figure 6 shows the range of dry matter contents and vane test cohesions encountered along with the (lack of) correlation between the two quantities. Figure 7 shows the correlation between the Indavertest and the vane shear test.

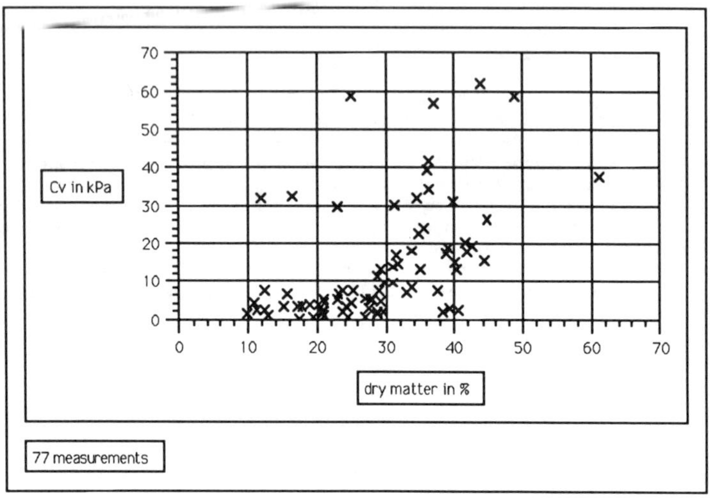

Figure 6 : Vane shear strength versus dry matter content.

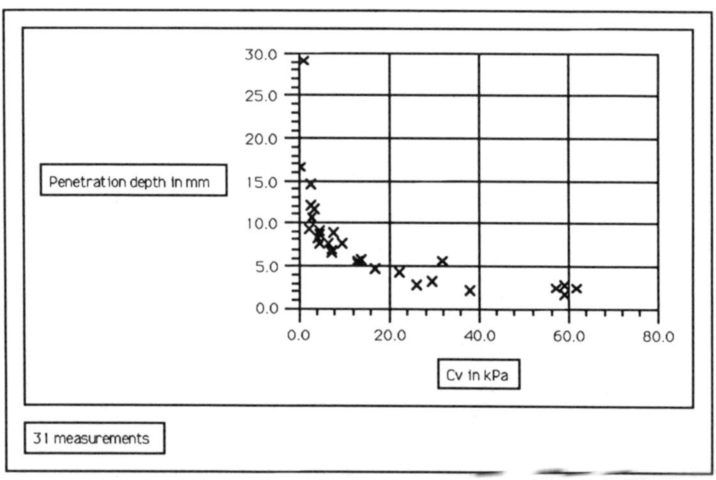

Figure 7 : Indavertest versus vane shear test.

Conclusion

Different routes may be followed to establish a standard for the mechanical strength as part of the landfilling requirements that must be met by sludge-like waste materials :

- the empirical route will correlate mechanical characteristics with problems as experienced by the landfilling exploitation;
- a theoretical route may be relied on to estimate stabilities and safety factors met in the application of sludges to landfills;
- looking at presently used or available dewatering techniques and available adding materials, the feasibility of any proposed criterium will have to be established;
- an eye must be kept on research and regulation in neighbouring countries and in the European context.

We hope to be able to follow these four routes in order to propose a standard that satisfies the technical, ecological as well as economical requirements imposed on landfilling of sludges. If such a standard would be generally accepted in the near future, it could take away much of the confusion that now reigns in this field

References

(1) P.Geuzens;
Vergelijkende studie betreffende bepalingsmethoden van slib-stabiliteit en rheologische eigenschappen;
SVW report 1989/1.

(2) Sewage Sludge Treatment and Use; (EUR 11918)
A.H.Dirkzwager & P.l'Hermite (ed.);
Elsevier, 1989;
(Proceedings of a COST 681 Conference held in Amsterdam, The Netherlands, 19-23 September 1988).

(3) Loll, Kassner, Köhlhoff, a.o.;
Die Bestimmung der Deponierfähigkeit von Schlämmen mit der Referenzmethode "Laborflügelscherfestigkeit";
in : Korrespondenz Abwasser, 8/89, p.903-908.

(4) B.Vanderborght & R.Ansoms;
Steekvastheid van industriële afvalstoffen;
in : Technivisie, nr.116, 1988.

PC-BASED OPERATOR ASSISTING SYSTEM FOR SLUDGE DEWATERING

H. GRÜTTNER AND J. LA COUR JANSEN,

Water Quality Institute
Science Park Hørsholm
11, Agern Allè
DK-2970 Hørsholm
Denmark

Summary:

On a municipal wastewater treatment plant using a decanter-centrifuge for dewatering of two anaerobic digested sludges with different characteristics a PC-based operator assisting system was developed. The system is based on a database used to collect data on sludge properties and operational conditions which is added by the operator. By graphical presentation and an advisory service these data are used to support the operator in his dewatering operations and to secure a running optimization of the sludge dewatering.

Introduction:

Sludge dewatering is an important part of the running operations at most wastewater treatment plants, due to several reasons:

1) When the sludge is deposited in a landfill or taken to an incinerator plant, the dry solids contents in the dewatered sludge greatly influence the economy in such treatments.

2) When not treated properly the sludge treatment and dewatering may give a considered amounts of dissolved or suspended matter which is returned to the water treatment part of the plant.

3) Often a significant amounts of the running cost is used on sludge treatment, dewatering and transportation - in Denmark 40% to 50% is typical.

At the same time the operation of the dewatering equipment is often not subject to a systematic optimization. This is due to the fact that many factors influence the dewatering operation, and in many cases the mecanisms are not fully understood neither by the operator nor the involved experts. One of the reasons for that is the fact that in many cases there is no systematic collection of data from the running operation.

In the light of this it was considered an important task to develop a system which would make it possible for the operator to collect his data in an easy way, and to give him a tool for using these data in his further optimization of the system.

Such a system was developed on one plant and has later been transferred to another plant for evaluation.
In an overall description the system consists of three main parts:

* Data on sludge proporties, operational conditions, results from dewatering and operator observations makes a database, which is the backbone of the system.

* Graphical presentation of data and observations from the database and collocations of data gives the operator immediately access to the past experience with the dewatering system.

* An advisory service is installed in order to support in case of operational problems or wishes for a specific optimization of the system.

The system, which is completely based on input from the operator, is presented in further details below together with some of the results obtained.

Dewatering plant

The system was developed at Sjælsø Wastewater Treatment Plant where it has been in operation since January 1990. This plant treats domestic sewage from about 30.000 PE from the town of Birkerød north of Copenhagen, Denmark. The plant concists of a conventional biological trickling filter and a activated sludge nitrification step.

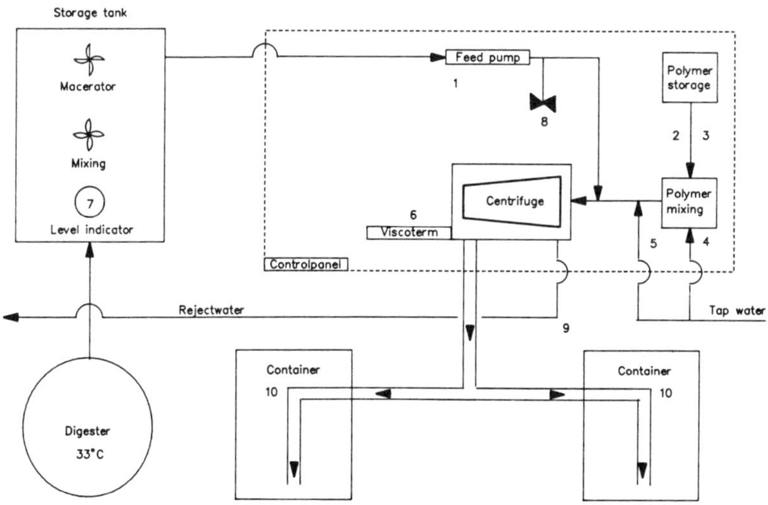

Figure 1 : Scematic diagram of the dewatering equipment at Sjælsø wastewater treatment plant.

Sludge from both systems is anaerobic digested together. The dewatering equipment is further used for treatment of sludge from a neighbouring plant (Bistrup). This sludge is anaerobic digested sludge from a conventional trickling filter.

A principal diagram for sludge dewatering at the Sjælsø plant is shown i figure 1 with informations on the tools for adjustments and readouts. Please note that there are no sensors for turbidity or flow in the system.

The centrifuge is a Decanter 142 C from DDS-Engineering, with a nominal capacity of 10 m3/h, and a maximum speed of 2780 min-1 resulting in a G-max of 1665 G. It is equipped with a hydraulic system for automatic backdrive torque control, called Viscotherm Rotodiff.

Structure of databases

The overall philosophy has been to make it as easy as possible for the operator to use the system. It means that the same data are only put into the system once, and during operation he only needs to put in new data if a parameter is changed. But at the same time it is tried to illustrate for him which parameters are relevant to register when adjusting the operation or taking a sample.

As stated above a separation is made between parameters which are constant for a longer time of operation (table 1), and those which may be changed during operation and hence need to be updated by each adjustment (table 2). A very important part of this database is the "reason for adjustment" which the operator is asked to fill out by each adjustment, either by choosing from a list of typical reasons, or he can create a new if he wants.

A third database stores the analytical data (table 3). The parameters in this list is - in contrast to the operational parameters - optional, so the operator can use which of them he wants. In addition to dry solids and temperature, Capilary suction time (CST) has been introduced. The CST is used to calculate the "sludge-factor" (CST/DS), as an easy way to test whether the sludge has changed.

The system is made in a way so that it can handle the fact that many adjustments are made without taking a sample, and in the situation where a sample is taken, the result will be registered some hours later.

As a service to the operator those of his registrations which are not in standard units are immediately transferred, e.g. the speed of the feed-pump (the Tachometer-value) is transferred to flow. In addition some calculations are made:

- the actual amount of polymer used
- the actual concentration of polymer added
- the polymer dosing (g/kg DS) - if dry solids is known.

The system is created in TURBO-PASCAL with the use of Database Toolbox.

Using the program

The system is menu-based with the following options in the main menu:

- Journal for operation
- Graphs
- Reports
- Evaluation of data

MASHINE
- water level in bowl

POLYMER
- type of polymer
- concentration of polymer
- polymerpump - length of strokes

SLUDGE
- type of sludge
- remarks

OPERATION
- start-up - time
- initial for operator

Table 1 : Data registered by start-up

ADJUSTMENTS
- level in storage tank (7)
- Tachometer (1)
- polymerpump - frequency (2)
- water for polymerdilution (4,5)
- oilpressure of Viscotherm (6)
- relative scroll speed

REASON FOR ADJUSTMENT
- standard reason or explanation

SAMPLE
- yes/no (yes generates number)

Table 2 : Data registered during operation

INLET SLUDGE
- temperature (°C)
- dry solids contents (%)
- Capilary suction time (sec.)

REJECT WATER
- temperature (°C)
- pH
- suspended solids (mg/l)

DEWATERED SLUDGE
- dry solids contents (%)

Table 3 : Analytical data registered

The "Journal of operation" is further structured in the following menu-points:

- Start operation
- Adjustments
- Samples
- End of operation

"Start operation" is used to register the basic informations mentioned in table 1, and the actual start-up values of all the parameters in table 2. It also asks the operator to identify himself, and to decribe the sludge he intents to work with.

In "Adjustments" he is asked why he made an adjustment, whether he wants to take a sample, and to update the parameters he has changed. If he answers that he will take a sample, the system generates room for a new sample in the database and gives him a number to identify the sample.

When he wants to put in his analytical results in "Samples", he just needs to identify the number of the sample. When the results are typed, the system calculates the polymerdosing (g/kg DS) and the solids removal (%) if the necessary results are available.

In "End of operation" the operator is asked to give figures on the hours of operation, the sludge left in storage-tank for the next day and the amounts of dewatered sludge transported from the plant. In return he gets a small report which summarises the amounts of sludge treated, the polymer used and the water used and averages the samples taken at that particular run.

The same figures are available on a monthly basis in the "Reports" in the main menu, together with some other reports. The reports may also be printed as a documentation of the operation of the dewatering equipment.

"Graphs" contains two types of graphs. The presentation of some of the in-data values and the calculated values on a time scale, and a few graphs where the values are set in relation e.g. the polymerdosing as a function of solids content.

The above mentioned parts of the system has been used for 5 month at the Sjælsø-plant and some of the results collected will be described below.

The final part of the system, "Evaluation of data" is still under preparation. It consists of two ways to extract data from the database:

1) The operator is asked to give priority to the different options: High Dry solids in dewatered sludge
Low suspended solids in reject-water
Low polymer-use
High capacity

Based on his priority, the data will be extracted and presented on the screen, so that he will be able to see how the equipment was adjusted in situations similar to his high-priority situation.

2) The operator may ask: "What did I do, when I had the problem ....", using his own list of reasons for adjustments. The system will then provide the data from similar situations in a way that will make him see which of his adjustments that had a benificial effect.

When this part of the system is put into operation it will make it possible to compare the two situations, and to see whether the advisery service as expected will provide an optimization of the system.

Results from operation

During the first four months of operation only data collection has been installed, as described above. The strategy for the running operation of the decanter has been to keep the dry solids in the dewatered sludge above 30% and to have a reasonable rejectwater quality.

The operators have used the system fairly often. As described in table 4, they have put in data three to four times a day on an average basis, in addition to the start-up procedure. Typically they put in one serie of analysis per day.

Sludge origin	Number of data	start-ups	adjustment	Adjustments per day
Sjælsø	155	37	118	3,1
Bistrup	41	8	33	4,1
Mix	11	2	9	4,5
Total	207	47	160	3,4

Table 4 : Numbers of data and adjustment for the first four months operation

On the other hand it seems that they have had little advantage from the system. As an example the control of polymerdosis will be described.

As it can be seen in figure 2 there is a need for a clear feed-back to the operator, since he obviosly keeps his polymer addition relatively constant. As a consequense the variations in the dry solids contents in the inlet, produces a very wide range of polymerdosis on a dry solids basis.

The above example is a very good illustration of the fact that the operator actually has the informations to make a reasonable control, but does not use it. In fact, the system calculates the polymerdosis on dry solids basis each time he puts in a value for dry solids in the inlet. In only one situation this information provoked an adjustment.

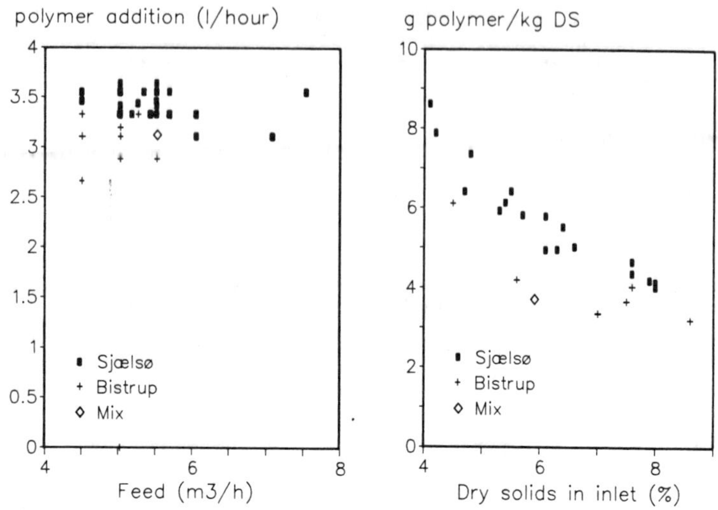

Figure 2 : Relations between polymer addition and feed, and between polymer dosing and dry solids in inlet.

The reasons for adjustments can be used to evaluate whether the operator had made a reasonable adjustment in a given situation. As an example the adjustments made to cure "low solids in dewatered sludge" are shown in table 5.

As it can be seen from the table his adjustments are not exactly systematic. The most reasonable thing to do - to lower the relatives - was only done in 4 out of 57 adjustments where he indicated that his wish was to get a higher solids content in the dewatered sludge.

	UP	DOWN	TOTAL
Feed flow	14	9	23
Polymer	13	5	18
Relatives	12	4	16

Table 5 : Adjustments made to get a higher solids concentration in the dewatered sludge

Discussion of the approach

The above mentioned example on polymer addition might of course be solved by adding a measuring device to the system, but that might on the other hand introduce a new problem in the operation: to test when to measurement is OK.

The general situation for many small wastewater treatment plants in Denmark, is that the people are present at the plants, but they need support and education. That means that introduction of an automatic controlsystem, described by several authors (e.g.: 1, 2, 3), may optimize the dewatering on a short range, but in the long run it is better to give the operators an understanding of the mecanisms controlling the processes they are working with.

We think that in the future there is a need for systems like the one described above, used as a tool for supporting the unskilled operator, guiding the skilled operator through further optimizations and used also as a way to tranfer knowledge from one operator to another. The systems for automatic control based on several on-line measurements, may be used on bigger plants to control the individual mashines, and may to our oppinion take over the control only within specific limits or be used as a consulting system giving information and advices to the operators.

With the price level on computerforce it is very likely that in the future each plant will have its own model for dewatering like the one described in (4). Such an empirical model might be used as a tool for giving advices to operators and the data collected will provide a continous calibration.

Acknowledgement

This project in supported by the National Agency for Environmental Protection and The Board of Technology.

References

(1) WIKSELL, H. and de BESCHE, S. (1978). Automatic control of a centrifuge process. Vatten 1, 10-14.

(2) CAMPBELL, H.W. and CRESCUOLO, P.J. (1988) Control of polymer addition for sludge conditioning: A demonstration study. in: What's new in wastewater technology?, Eds: Environment Canada and Canadian Water and Wastewater Association. 25-40.

(3) WINTER, F. (1990). Optimierte Flokkungsmitteldosierung am Beispiel der Schlammeindickung/Schlammmentwässerung. Korrespondenz Abwasser 4, 296-392.

(3) HERATH, B., GELADI,P. and ALBANO, C. (1989). Developing an empirical model for dewatering... Filtration & Separation jan/feb, 53-56.

INFLUENCE OF DIGESTION ON SLUDGE CHARACTERISTICS

V. LOTITO*, L. SPINOSA* and M. SANTORI**

*CNR-IRSA, Via De Blasio 5, 70123-BARI (I)
**CNR-IRSA, Via Reno 1, 00198-ROMA (I)

Summary

Research was carried out at CNR-Istituto di Ricerca Sulle Acque (CNR-IRSA) to investigate the influence of aerobic, anaerobic and chemical digestion processes on sewage sludge stability, dewaterability and hygienization. A stable sludge, as measured by volatile total solids concentration, was obtained after about 15 days of aerobic digestion and 25 days of anaerobic one; for chemical stabilization the volatile solids reduction was obtained as a mass effect of lime addition. Results showed that the influence of digestion time on dewaterability, as measured by specific resistance to filtration and compressibility factor, can be neglected because its effect is scarce if compared to that of chemical conditioning. As far as hygienization is concerned, aerobic and anaerobic digestion times longer than those suitable for stabilization purposes are needed to obtain a pathogen reduction by one order of magnitude at least. High lime dosages allow pathogens to be reduced to a larger extent.

1. INTRODUCTION

Among the processes which sewage sludges can be submitted to, stabilization plays an important role as it is generally performed and influences many sludge characteristics. However, information available in literature on this subject is not too much abundant and quite discordant.

Therefore, research on laboratory scale was carried out at CNR-IRSA in order to investigate the influence of stabilization processes, i.e. aerobic, anaerobic and chemical, on sludge stability, dewaterability and hygienization. The above sludge characteristics were selected because they are representative of very important aspects involved in sewage sludge management. Stability is generally associated with putrescibility, or the tendency of organic matter to biodegrade, thus leading to odour production. Dewatering is almost always adopted in sludge processing because it allows volume, and therefore subsequent treatment and disposal costs, to be consistently reduced. Hygienic characteristics are very important because pathogens can pose potential threats to the public health and environment during disposal operations.

This research follows a preliminary one carried out by CNR-IRSA on the influence of aerobic and anaerobic digestion time on sewage sludge stability and dewaterability (1).

2. BACKGROUND

As for stability, volatile suspended solids destruction appeared to be a simple and reliable parameter for its assessment. A destruction of 30-40% was reached after about 15 days of aerobic digestion or 25-30 days of anaerobic one (1).

Data of volatile solids destruction vs. time deriving from pilot and full scale tests of several types of sewage sludge indicated that reductions between 25 and 40% can be obtained in 10 to 30 days (2).

D'Antonio (3) suggested that total suspended solids, rather than volatile suspended solids or degradable suspended solids, is the parameter best fitting the experimental results in aerobic stabilization.

In anaerobic digestion a volatile suspended solids destruction of 49.0 to 62.7% for primary sludge and of 40.3 to 49.9 for activated one was found by Chapman (4), of 60% for primary sludge and of 40% for activated one, depending on sludge solubility, chemical composition and structure, by Stuckey (5) and of 54±17% for mixed sludges, deriving from 30 different domestic wastewater treatment plants, by Speece (6).

Kugelman and Jeris (7) reported that digested sludge from most full-scale digesters contains 45-50 % of volatile solids.

The stabilization of sludges with chemicals may have some advantages over biological methods in terms of capital cost, simplicity of operation and immediacy of the effect. Disadvantages include high operating cost, the need of close control and a temporary stabilizing effect in the case of lime. The process entails the addition of sufficient lime (usually in hydrated form) to raise the sludge pH to 12 and maintain it at about that level for some days. The main effect of a high pH is to suppress the emission of volatile sulfides and fatty acids, while that of amines and ammonia will increase. Usual dosages are in the range 150-200 $gCa(OH)_2/kg$ dry solids (8).

Other experiences indicated that significant pH decay should not occur once sufficient lime has been added to raise the pH to 12.5 and maintain that value 30 minutes at least. The actual volatile solids concentration following lime stabilization are lower than those which would result only from the addition of lime, as a result of neutralization, saponification and hydrolisis reactions which convert solids into soluble forms (9).

As far as the effect of digestion on dewaterability is concerned, previous tests indicated that specific resistance to filtration is slightly reduced by aerobic digestion, but this advantage is partially cancelled by the opposite behaviour of compressibility. Just the opposite occurred in anaerobic digestion tests. In any case, digestion time did not show any significant effect on specific resistance, if compared to that of conditioning, and probably not generally applicable to all the sludges to the same extent (1).

It was also found that specific resistance is reduced by 10-50% in 1 to 5 days of aerobic stabilization, but then it increses in the following 11 to 15 days to values higher than the initial ones (10).

According to other experiments, aerobic stabilization was found to have a negative effect during the first days of aeration (3). In any case specific resistance remained in the same order of magnitude.

Dewaterability appeared to be reduced by anaerobic conditions as indicated by the fact that the cost of digested sludge conditioning was about double (11). The same was observed in laboratory scale tests by Chapman (4): anaerobic digestion sharply increased activated sludge specific resistance, while had not any marked influence on that of primary sludge.

Lime addition in dosages sufficient for stabilization do not give proper conditioning effect for dewatering. Specific resistance of lime stabilized sludge generally improved with lime addition, but during the storage period before dewatering filterability could deteriorate (12).

Lime showed to have some slight dehydration effects on colloids; however it provides a granular structure which increases sludge porosity and reduces sludge compressibility (2).

Removal of pathogen is not usually the principal aim of stabilization, but most processes either passively allow pathogens to become non-infective or actively disinfect, as far as total coliforms, streptococci and salmonellas are concerned.

Aerobic digestion tipically reduces viral and bacterial pathogens by one order of magnitude (13, 14).

Higher reduction up two orders of magnitude could be obtained by anaerobic digestion (2, 13, 15).

In general, the effectiveness of lime stabilization in controlling pathogens depends on maintaining the pH at levels that destroy microorganisms and inhibit growth, should contamination occur after treatment. Reductions up to four orders of magnitude are reported in literature (2).

3. EXPERIMENTAL

Different municipal sewage sludges, whose characteristics are reported in Table 1, were used for experiments.

Table 1 - Characteristics of tested sludges

Sludge	SS (mg/l)	VS/TS (%)	R ($m/kg*10^{14}$)	S
a)	18,100	72.4	0.18	0.91
b)	20,500	73.5	1.77	0.87
c)	26,700	71.8	0.56	0.98

Aerobic digestion tests were carried out in 3 periods lasting 14 days the first and 28 days the others. A tank of 80 l aerated by porous diffusers was used; air was supplied by a compressor and controlled by maintaining the dissolved oxygen level around 2 mg/l. During each run, temperature (T), total solids (TS), volatile total solids (VS), volatile to total solids ratio (VS/TS), specific resistance to filtration (R), compressibility factor (S), total and fecal coliforms, fecal streptococci and salmonellas were measured. The analytical standard methods for sludge characterization by CNR-IRSA (16) were adopted.

The anaerobic digestion tests were conducted in a laboratory apparatus consisting of several eudiometric tubes (700 ml) each connected to a bottle (500 ml) immersed in a thermostatic bath maintained at 37 °C. The bottles were filled with the sludge to be tested without any seed in order to avoid possible effects induced by it, expecially on pathogens. Biogas production was detected every day, while the other parameters above listed were determined at the beginning and the end of the digestion time. Duration of tests was of 20, 30 and 55 days, respectively.

Chemical stabilization was performed in 10 l tanks by lime addition. Two lime dosages were adopted: the first (I) able to reach a pH of 12 and maintain it for 2 h; the second (II) exactly the double. Sludge was mixed for two hours in the first day and for half an hour each day before sampling. Three runs were conducted at the following (I) dosages: 166, 216 and 220 kg/t TS.

4. RESULTS

In the following, results obtained in the experimental work on sludge stability, dewaterability and hygienization are respectively discussed.

Stability

VS destruction was confirmed to be an useful parameter for assessing sludge stability.

A destruction of 20% at least, starting from a VS/TS ratio of 71.8 to 73.5, was reached after 15 days of aerobic digestion, but values above 30-35%, coupled to significant reduction of pathogens content, were obtained only after 30 days of stabilization.

The same occurred for anaerobic digestion, also due to the absence of seed sludge, but digestion times 1.5 longer than the previous ones were required.

In chemical stabilization a reduction of VS/TS was observed, obviously mainly due to the mass effect of lime addition, and, therefore, to its dosage. It was of about 20% at I dosage and 30% at II.

Dewaterability

R and S were used as characterizing parameters of sludge dewaterability.

During aerobic stabilization R appeared to remain quite constant; if some changes occur, they can be normally observed during the first 2-3 days of processing. Generally speaking, R and S tended to be modified in opposite way, so any effect on R is partially counterbalanced by the opposite behaviour of S (Fig. 1). In any case, changes were negligible if compared to those obtained by conditioning.

As shown in Fig. 2, R tended to be slightly increased by anaerobic digestion, while S was more consistently reduced by 25-50% with time of digestion; this can be explained in terms of particle size reduction.

Chemical stabilization generally caused a sudden increasing of R up to one order of magnitude, which remained quite constant with time, even if at higher dosages a certain further reduction was evidenced. This could depend on lysis of cellular material due to the action of lime. On the contrary, S tended to be strongly reduced by chemical stabilization, expecially at higher dosages, thus showing the tendency of the sludge to become not compressible, as a consequence of floc structure modification (Figs. 3 and 4).

Hygienization

Results are summarized in Table 2.

Table 2 - Effect of stabilization processes on pathogens

Sludge	Process	Tot. Col. (MPN/100ml)	Fec. Col. (MPN/100ml)	Fec. Strept. (MPN/100ml)
	Raw	$7.5*10^6$	$1.1*10^6$	$7.5*10^6$
	Aerobic 14d	$2.3*10^6$	$1.1*10^6$	$1.1*10^6$
a)	Anaer. 20d	$2.3*10^6$	$4.0*10^5$	$2.3*10^6$
	Chem. I 1d	$2.1*10^6$	$1.1*10^4$	$2.1*10^6$
	Chem. II 1d	$9.3*10^4$	$7.0*10^3$	$1.5*10^4$
	Raw	$2.1*10^6$	$2.3*10^4$	$1.1*10^6$
	Aerobic 28d	$4.6*10^5$	$9.0*10^3$	$1.5*10^5$
b)	Anaer. 30d	$2.1*10^5$	$2.3*10^4$	$1.1*10^5$
	Chem. I 1d	$2.4*10^5$	$1.5*10^4$	$2.4*10^5$
	Chem. II 1d	$2.1*10^4$	$4.0*10^3$	$2.1*10^4$
	Raw	$1.1*10^8$	$2.3*10^6$	$1.1*10^7$
	Aerobic 28d	$2.4*10^5$	$2.3*10^4$	$2.3*10^4$
c)	Anaer. 55d	$4.6*10^4$	$4.0*10^3$	$4.0*10^3$
	Chem. I 1d	$4.0*10^5$	$2.3*10^4$	$2.4*10^5$
	Chem. II 1d	$2.0*10^6$	$4.0*10^3$	$4.0*10^5$

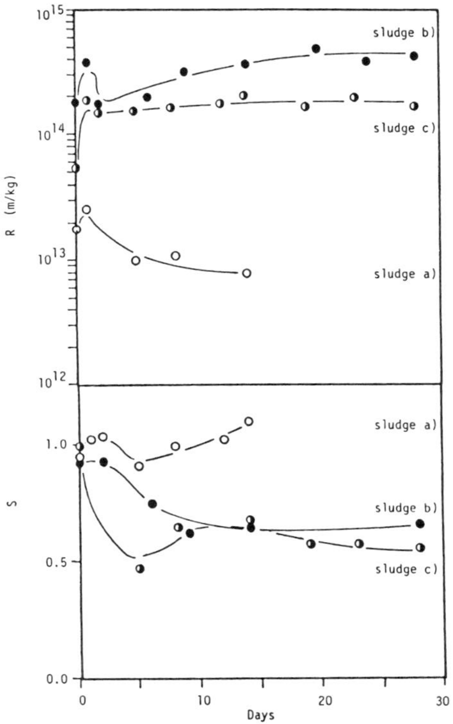

Fig 1 - Effect of aerobic stabilization on dewaterability

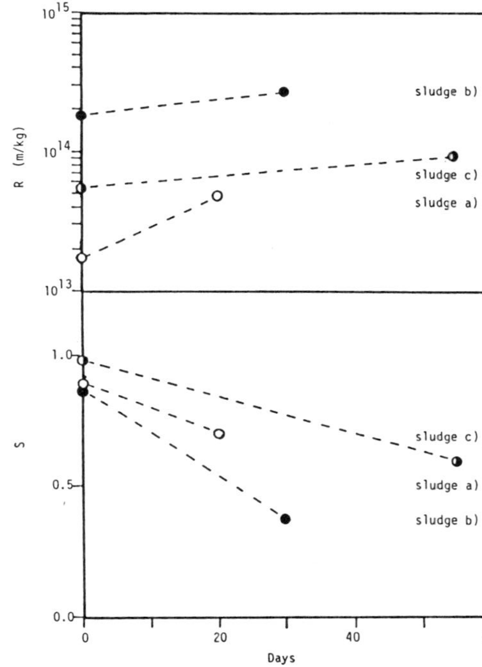

Fig 2 - Effect of anaerobic digestion on dewaterability

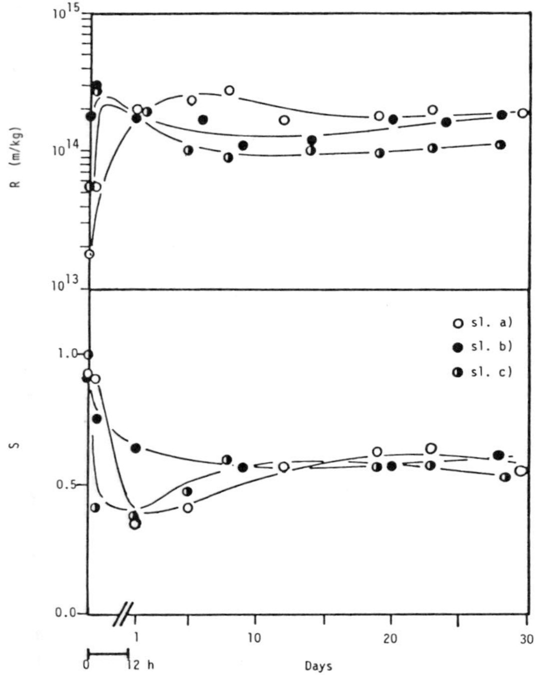

Fig 3 - Effect of chemical stabilization (dos. I) on dewaterability

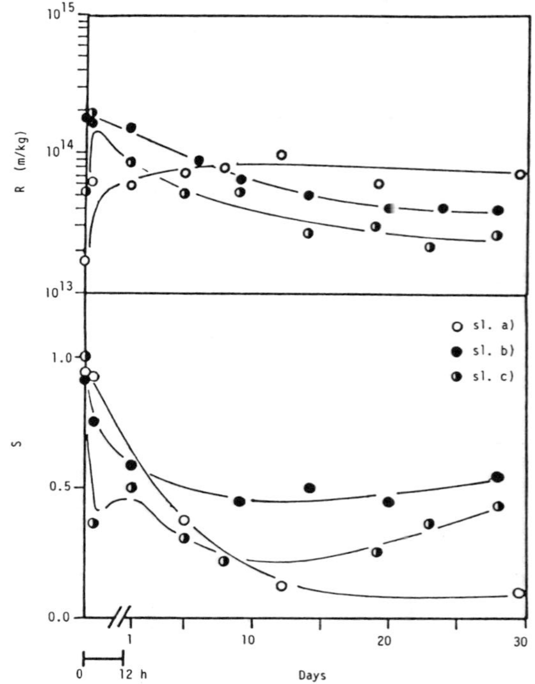

Fig 4 - Effect of chemical stabilization (dos. II) on dewaterability

Effects were poor after 14 days of aerobic stabilization, 20 days of anaerobic digestion and at I dosage of chemical digestion. Only with II dosage results were good, in that reduction of two orders of magnitude for total coliforms and fecal streptococci and three orders for fecal coliforms were obtained. Prolonging both aerobic and anaerobic digestion to 28 and 30 days respectively allowed a certain degree of hygienization to be obtained, by reducing pathogens of one order of magnitude.

Further better results were obtained by amelioring the process conditions in aerobic digestion at 28 days, by prolonging anaerobic digestion to 55 days and improving lime mixing in chemical conditioning, expecially at lower dosage.

5. CONCLUSIONS

Sewage sludge stability was assessed by volatile solids destruction. Values higher than 30%, coupled to one order of magnitude of pathogens reduction at least, were obtained after 28 days of aerobic stabilization and 30 days of anaerobic digestion. Longer times and/or better process conditions allowed higher hygienization effects to be obtained.

In chemical stabilization volatile solids abatement was observed as a mass effect of lime addition, but reduction of pathogens content of two to three orders of magnitude were obtained at higher dosages.

Results showed that the influence of stabilization processes on dewaterability can be neglected if compared to that of conditioning. Generally speaking, to any variation of specific resistance to filtration corresponds an opposite change of compressibility factor. In chemical stabilization a tendency to incompressibility can be evidenced, proportional to lime dosage.

Aknowledgments

Thanks are due to G. Barile and F. Lorè for their valuable contribution during the experimental work.

REFERENCES

(1) LI YAN, BARILE, G., LORE', F., LOTITO, V. and SPINOSA, L. (1987). Influence of digestion on sewage sludge stability and dewaterability. Environmental Technology Letters, 8, 249-259.

(2) U.S.EPA (1979). Report 625/1-79-011.

(3) D'ANTONIO, G. (1983). Aerobic digestion of thickened activated sludge- Reaction rate constant determination and process performance. Water Research, 17, 1525-1531.

(4) CHAPMAN, W.H. (1965). Some experiments on sludge digestion. Journal Inst. Sew. Purification, 485-493.

(5) STUCKEY, D.C. (1984). Assessment of anaerobic digestability of sludge. In: Methods of characterization of sewage sludge. D. Reidel Publ. Co., Dordrecht (NL), 123-130.

(6) SPEECE, D.C. (1983). Private communication.

(7) KUGELMANN, I.J. and JERIS, J.S. (1981). Anaerobic digestion. In: Sludge treatment. W.W. Jr Eckenfelder and

C.J. Santhanam Eds., Marcell Dekker Inc., New York, Caph 7, 211-276.

(8) BRUCE, A.M. and FISHER, W.J. (1984). Sludge stabilization-Methods and measurements. In: Sewage sludge stabilization and disinfection. A. Bruce Ed., Ellis Horwood Ltd, Chichester (U.K.), Chap. 1, 23-47.

(9) U.S.EPA (1978). Report 600/2-78-171.

(10) RANDALL, C.W., PARKER, D.G. and RIVERA-CORDERO, A. (1973). Bulletin 61. Virginia Water Resources Research Center, Virginia Polytechnic Institute and State University, Blacksburg, 86 pp.

(11) DICK, R.I. and HASIT, Y. (1981). Project Report, Grant No. ENV 77-22947, National Science Foundation, 170 pp.

(12) PAULSRUD, B. and EIKUM, A.S. (1984). Experiences with lime stabilization and composting of sewage sludge. In: Sewage sludge stabilization and disinfection. A. Bruce Ed. Ellis Horwood Ltd., Chichester (U.K.), Chap. 13, 261-277.

(13) U.S.EPA (1989). Report 625/10-89/006.

(14) GENEVINI, P.L., VISMARA, R. and MEZZANOTTE, V. (1983). Utilizzo agricolo dei fanghi di depurazione. Ingegneria Ambientale 12/9.

(15) HUDSON, J.A., BRUCE, A.M., OLIVER, B. and AUTY, D. (1987). Operating experiences of sludge disinfection and stabilisation at Colburn sewage works. Meeting of the Institution of Water and Environmental Management, Tyne & Humber Branch, WRC Report, Stevenage (U.K.), 21 pp.

(16) CNR-IRSA (1984). Metodi Analitici per i fanghi. Report 64.

SESSION I

GENERAL DISCUSSION

REGARDING PAPER 1

J E HALL

I was very interested in Dr Colin's paper as his studies are similar to some work carried out a few years ago at WRc in the UK. I would like to ask why in Dr Colin's studies, the fresh sludge is mixed first with old domestic waste?

F COLIN

That is to isolate the effect of the fresh sludge since the old domestic waste would not have any effect.

L LANGGAARD

What is the pH in the column?

F COLIN

That depends on the experimental conditions. Initially the pH is about 7. However when organic wastes are added to the column the pH drops by about 0.5-1.5 pH units. With fresh refuse, the pH in the column falls to about 5.

P GEUZENS

Can you predict the quality of the percolate water from your experiments?

F COLIN

That is difficult because the nature of waste refuse varies so much. Attempts have been made to model landfills to predict the quality of waste water evolution. These, however, only work well with simple sludges and wastes.

REGARDING PAPER 2

J J VAN DEN BERG

Have you considered replacing or supplementing the shear-strength test?

P GEUZENS

It is only an empirical test and not very fundamental. It can, however, be useful in predicting sludge characteristics.

L SPINOSA

What is the weight of the penetrometer used and has the limit of the shear strength been determined?

P GEUZENS

The penetrometer weighs 100g. As yet, the limit of the shear strength has not been determined.

J E HALL

Do you expect the proportion of sewage sludge disposed to agricultural land to decrease markedly over the next 5-10 years?

P GEUZENS

That could happen because of the increasingly tighter controls on the use of sewage sludge in agriculture being introduced and the natural concern about the practice expressed by farmers. In that case, more sludge will have to be disposed of by other routes, for example, by landfilling or incineration.

REGARDING PAPER 3

H CAMPBELL

I would like to congratulate the authors on their paper. A paper outlining the practical considerations of running a wastewater treatment plant is long overdue. Parameters suitable to be used by plant operators have not usually been defined. More priority tends to be given by the plant management to the automatic control of the plant. In that case, control tends to approximate to a pass/fail test. This tends to be limited by the availability of sensors for, as an example, cake solids.

H GRÜTTNER

I agree with Mr Campbell's remarks. In general I think it better to control the plant via the solids content of the sludge rather than the effluent quality.

L SPINOSA

Do you use the CST apparatus to optimise the operation of the dewatering plant?

H GRÜTTNER

We did not use it in our work though the CST apparatus is a useful means to obtain a characterisation of the general quality of the sludge.

REGARDING PAPER 4

J E HALL

Do you feed your digesters batch-wise or continuously?

L SPINOSA

They are fed batch-wise. This explains the initial drop in the faecal coliform content. With raw waste being fed to the digester it is difficult to get the reaction to start.

B PAULSRUD

That it a similar experience to that observed with wet sludges where it is preferable to treat them first aerobically.

J E HALL

As the solids content in the digester is as high as 23% dry solids, what mixing do you employ?

L SPINOSA

In general the solids content is not as high as 23% dry solids and the outflow material is usually about 10% dry solids. Some mixing may be necessary though the bio-gas production will help.

L LANGGAARD

Have you calculated the cost of digestion, expressed in terms of cost/m^3 of sludge treated, of the various treatment options that you considered.

L SPINOSA

Detailed costings have not been undertaken. However, it is likely to be somewhere between 100 and 200 ECU/m^3 sludge treated.

Session II

DISPOSAL STUDIES

Chairman : F. COLIN

Rapporteur : P.J. NEWMAN

SOME ECONOMIC AND QUALITY ASPECTS OF COMPOSTING SEWAGE SLUDGE

S.R. SMITH and J.E. HALL
Water Research Centre, Henley Road, Medmenham, Marlow, Buckinghamshire SL7 2HD, UK

Summary

Composting in the UK is seen increasingly as a potentially viable treatment method for many wastes, including sewage sludge. This paper describes part of a three year research programme which has had the overall objectives of optimising methods of composting sewage sludge with cereal straw, and of assessing the operational, marketing and economic implications of the process and the product. Two methods of composting dewatered polyelectrolyte conditioned sewage sludge with straw are described. Aerated static pile and turned windrow systems were developed to produce a bulky organic material suitable for use in agriculture as a soil conditioner and slow-release fertiliser. Temperature conditions during composting fully satisfied the UK Department of the Environment Code of Practice requirements for sludge treatment. A significant bulk reduction occurred during composting through moisture and volatile solids loss. This resulted in a concomitant increase in compost heavy metal concentrations. Effects of composting on metal content are discussed in relation to the potential environmental and economic benefits of composting as a sludge management option.

1. INTRODUCTION

Composting can offer several potential benefits in the treatment and disposal of sewage sludge. Transport costs may be reduced due to volume and weight reduction through the loss of volatile solids and moisture. The composted material is stabilised and more aesthetically acceptable with improved handling and storage characteristics compared with raw sludge. It is, therefore, more amenable to disposal or utilisation. Composting reduces the emission of offensive odours at the sewage treatment works. Odour emissions from compost applied to agricultural land are also reduced compared with untreated sewage sludge. In addition, pathogenic organisms are inactivated due to the elevated temperature conditions which occur during composting. Finally, the composted end-product may have a market value as a soil conditioner and fertiliser for use in agriculture, amenity and tree planting.

Sewage sludge will not compost directly and requires processing with a bulking agent to improve the aeration characteristics of the sludge. Bulking agent addition also encourages effective composting by increasing the dry solids content and carbon-nitrogen ratio of the feedstock mix. Under UK conditions, cereal straw is the most readily available and inexpensive material with suitable structural characteristics for use as a bulking agent.

Much research has been carried out to develop process technology to improve efficiency of composting sewage sludge primarily by aerated static pile and turned windrow methods (1-8). With static pile systems there is no further mixing of the sludge and bulking agent after the initial blending until the material is removed from the aeration manifold. Evaporative heat loss by forced aeration maintains compost temperature at optimum levels for organic matter stabilisation by preventing the excessive accumulation of metabolically generated heat (4). In contrast, turned windrows are agitated frequently but temperature control is limited compared with that achieved by forced aeration systems operated by temperature feedback (1). Biological aspects of composting including microbial population diversity and rate of activity, pathogen kill, and nitrogen transformations are also well reported (9-17). Sewage sludge will usually contain contaminants such as heavy metals, particularly when derived from industrial areas. However, few quantitative data are available demonstrating effects of composting on these sludge contaminants.

One of the aims of this study was to determine the effect of composting on the dynamics of changing heavy metal concentrations in sewage sludge-straw compost produced by aerated static pile and turned windrow methods. The results are discussed in relation to the suitability of composting as a sludge management option.

2. MATERIALS AND METHODS

2.1 Static Pile Composting

2.1.1 Aeration Manifold Design and Process Control System

The aeration manifold system consisted of a galvanised steel duct pipe (33 m long; 0.6 m diameter) with 31 outlet ports connected to capped perforated lateral pipes (6 m long). Dual stage fans were installed at each end of the main duct. An air operated damper was fitted at one end of the duct, opposite the first two fan stages. The damper prevented air passing directly through the main duct when these fans operated, thus ensuring air flowed through the lateral pipes.

The fans were operated by a microprocessor which also logged temperature data from 16 thermocouples inserted in the composting mass. The process control strategy was similar to that proposed by Finstein *et al* (4). The computer was programmed so that one fan stage was actuated when the mean compost temperature increased above 56°C. This aeration regime was maintained until compost temperature decreased below 55°C when the fan was switched off. However, if the temperature continued to rise an additional fan stage came into operation at 60°C. All four fans operated at temperatures greater than 65°C. In addition to the temperature feedback control, a time-based default system switched on a single fan stage for three minutes at 30 minute intervals. This ensured oxygen did not become limiting when the temperature feedback mechanism was not operating.

2.1.2 Operational Procedures

Dewatered polyelectrolyte conditioned sewage sludge cake (dry matter content 21-25%) and wheat straw (dry matter content 80-90%) were combined by processing twice through a rear discharge muckspreader. Sludge and straw were processed to produce a feedstock material containing 7% straw (fresh weight, corresponding to 22% straw on a dry weight basis). The feedstock material was loaded onto the aeration manifold with a front-end loader on 25 August 1989.

Following completion of the aeration phase (24 days), the lateral pipes were removed and composted material was processed through the muckspreader. The material was reprocessed after a further 29 days maturation.

2.1.3 Mass Balance and Volume Loss

A mass balance was calculated by weighing the feedstock material over a weighbridge prior to loading on the manifold. Composted material was reweighed after stripping off the manifold and after the second reprocessing event. The volume loss was calculated on the basis of the number of loads weighed and the volume of the lorry tipper body.

2.2 Turned Windrow Composting

Windrows were constructed using a rear discharge muckspreader on 4 October 1989. Sludge and wheat straw was initially processed twice through the muckspreader to produce starting materials containing 5, 7 and 8.5% straw (fresh weight; corresponding to 17, 22 and 26% on a dry weight basis). Windrows were subsequently turned on four occasions at 7-10 day intervals and then once after a further 26 day period.

2.3 Temperature Measurement

The temperatures of blended sludge and straw were measured during the aeration phase of static pile composting using the system described earlier. In addition, temperature was measured for a further 65 days after removal of material from the aeration manifold, during the maturation phase. Temperature probes were constructed for this purpose using type K thermocouples. The probes were inserted into the material and left *in situ*. Three replicate measurements were taken at depths of 0.5, 1.0 and 1.5m. Windrow temperatures were measured at the centre of each pile with a portable electronic thermometer.

2.4 Sampling and Analysis

Sludge and compost samples were collected for analysis by bulking 10 subsamples. Sludge samples were collected periodically whilst the straw and sludge cake were being processed together. Static pile compost was sampled after the material was stripped from the manifold and remixed. Samples were also taken after the second reprocessing event and after 146 days maturation. The windrow was sampled 20 days after construction and also after 54 days following the final remix. The windrow was sampled again 106 days after the sludge and straw were initially combined. Bulked samples were thoroughly mixed and oven dried at 105°C and ground with a hammermill. Sludge and compost volatile solids were determined from the loss on ignition at 550°C. Compost bulk density was determined as the mass of dry compost per unit bulk volume, including the air space (18).

Oven dried and ground material was digested with nitric acid for heavy metal determination (Cu, Zn and Cd) using an atomic absorption spectrophotometer. Results of compost and sludge analyses are expressed on a dry weight basis. Values for initial feedstock mixtures were determined by calculation from analyses of the composite materials.

2.5 Statistical Treatment of Data

Linear regression procedures were applied to the data describing the relationship between compost volatile solids content and heavy metal concentration.

3. RESULTS

3.1 Qualitative Assessment of Compost Physical Characteristics

Sludge and straw at the surface of static piles retained their original characteristics during the aeration phase. However, uncomposted sludge and straw was soon degraded during the maturation phase when this material was incorporated into the bulk of the pile by processing through the muckspreader. In contrast, sludge and straw were integrally combined quickly by turned windrow composting due to the more frequent reprocessing regime.

Compost produced by both methods had an open, relatively friable texture, was dark brown in colour and had an earthy odour thus bearing no resemblance to the initial feedstock components.

3.2 Temperature Profiles

3.2.1 Static Pile Compost

Mean daily temperature of static pile compost was recorded during the aeration phase and is shown in Figure 1. In addition, daily maximum and minimum temperatures are presented. These were calculated as the mean of the three highest and lowest temperature readings recorded. Threshold temperatures given by the UK Department of the Environment Code of Practice process description for treating sludge by composting (19) are also shown. The Code of Practice requires that the temperature conditions should attain 55°C or above for at least four hours and over 40°C for a minimum of five days.

Mean daily compost temperature rapidly increased to 55°C and was maintained at this level for 16 days. Compost temperature declined over the remaining duration of the aeration phase. Maximum temperature increased to 65-67°C and minimum temperatures were in the range 32-35°C. When examined on an hourly basis (Figure 2), results indicated temperature increased up to the feedback set-point of 56°C and decreased to 53-54°C in response to forced aeration cooling. Compost temperature subsequently increased when the fan was switched off. Temperature feedback control was operating for a period of 350 hours.

Figure 3 shows the temperature of maturing compost after stripping off the manifold and processing through the muckspreader. Compost temperature increased after each reprocessing event and then subsequently declined.

3.2.2 Turned Windrow Compost

The temperature profile of windrowed compost is presented in Figure 4. Increasing the rate of straw addition in the feedstock material raised compost temperature. However, results indicate that temperature of compost containing 8.5% straw declined to lower levels after 40 days compared with smaller rates of straw addition. Reprocessing compost through the muckspreader significantly reduced windrow temperature. For example, on Day 38 remixing reduced compost temperature by 20-25°C. However, temperature subsequently increased following material agitation, but the rate of recovery was slower following successive reprocessing events.

3.3 Dry Matter and Volatile Solids Content

Compost dry matter content increased with time, whereas the level of volatile solids decreased for both composting systems examined (Figures 5 and 6). Increasing the rate of straw incorporated in windrowed compost generally raised the dry matter and volatile solids content of the material. Dry matter content of compost produced by both systems increased to a maximum of 35-40%. However, dry matter content of static pile compost declined after 146 days maturation due to winter rainfall.

3.4 Mass Balance and Volume Loss of Static Pile Compost

A mass balance for static pile compost is presented in Table 1 and volumetric losses are given in Table 2. The total mass of material was reduced by over a third after 24 days residence time on the manifold, which was associated with a 49% reduction in total volume. After 53 days, however, weight loss had increased such that less than half the mass of fresh material initially loaded on the manifold remained. A 59% reduction in total compost volume occurred over this period. Results indicated that there was a 21% loss of volatile solids during the aeration phase. There was a further small loss of volatile solids during the maturation phase such that the total loss was approximately 26% after the second remixing event. This was also reflected in a 56% reduction in the volume of dry solids during the aeration phase. However, there was little further change in volume during the subsequent maturation phase. In addition to moisture and dry solids loss, the volume of composting materials also declined through settlement which increased compost bulk density (Table 2).

3.5 Heavy Metal Status

Copper, zinc and cadmium concentrations increased during composting and were found to be inversely proportional to compost volatile solids content. Thus concentrations of metals in compost produced by aerated static pile and turned windrow methods are presented in Figures 7 and 8 respectively as a simple linear function of compost volatile solids content. Initial sludge metal concentrations are also shown.

The concentration of heavy metals in sludge-straw feedstocks were reduced compared with those in the sludge cake (Table 3). In addition, increasing the rate of straw addition reduced feedstock metal concentrations relative to those in the parent sludge (Figure 8). However, concentrations of metals in the blended material subsequently increased above those in the sludge as the volatile solids content declined during composting, by both static pile and turned windrow methods. Results indicated that a 20-40% increase in metal content may occur in the mature compost relative to the parent sludge. This corresponded to a 55-75% increase in metal concentration compared with that in the inital sludge-straw feedstock mix. Volatile solids loss increased metal concentration to similar levels in mature turned windrow compost irrespective of the initial rate of straw incorporation.

4. DISCUSSION

Dewatered, polyelectrolyte conditioned sewage sludge can be successfully composted with straw to produce a bulky organic material that is acceptable for use in agricultural crop production and amenity planting situations as a soil conditioner and slow release nutrient source (Smith, unpublished data). The material also has significantly improved odour, physical and handling characteristics compared with the raw sludge cake. Temperature conditions achieved during composting by aerated static pile and turned windrow methods fully satisfied the UK Department of the Environment Code of Practice requirements (19) for pathogen kill and organic matter stabilisation.

Water vaporisation is a major route of heat loss from composting materials and this is exploited in forced aeration systems controlled by temperature feedback mechanisms (5). Water vaporisation also explains why dry matter content increased during composting. However, dry matter content may be modulated by incident rainfall particularly during the maturation phase, unless conducted under cover, but the additional building costs would adversely affect composting economics.

The loss of water from composting materials significantly reduced compost mass and volume. Results reported here indicate that a 50% reduction in bulk of material composted by aerated static pile may occur. This clearly has major cost saving implications in compost transportation compared with uncomposted sludge cake. However, the mass reductions could not be fully explained in terms of increased dry matter content alone. This is because carbon dioxide, water and ammonia are also released through the mineralisation of organic matter (12) such that the volatile solids content of composting materials also declines. The dry matter and volatile solids content of turned windrow compost indicated that similar levels of bulk reduction to those obtained by aerated static pile composting may also be expected with this technique.

The loss of volatile solids during composting has important effects on the concentration of inorganic components, such as heavy metals, in the composted residue. Addition of the straw bulking agent initially reduced heavy metal concentration relative to the parent sludge, but decreasing compost volatile solids content raised the concentration of these components in the stabilised residue compared with that in the sludge cake. A similar or greater increase in heavy metal concentrations is observed during anaerobic digestion of sludge where volatile solids loss is generally around 40%.

The cost-effectiveness of composting dewatered sewage sludge cake as a method of sludge treatment is enhanced by maximising the rate of sludge treated such that the optimum quantity of straw used is comparatively small. In addition, the composting process may become moisture limited if the rate of straw addition is large (Smith, unpublished data). Therefore, little potential exists to reduce dewatered sludge cake compost metal levels through increased rate of straw addition. This indicates that the quality of sludge cake used for composting is important particularly if the residue is to be utilised for specialised horticultural purposes where the compost may form a significant proportion of the growing media. The level of sludge contamination is less critical if the stabilised material is to be used in general agriculture since suitable control measures are already available (19). Increasing the concentration of heavy metals by composting has to be balanced against the potential environmental and economic benefits which composting provides in the treatment and disposal of sewage sludge.

Composting has a number of potential advantages over alternative methods of sludge treatment. Bulk reduction reduces transport costs, stabilisation reduces odour emissions and the improvement to the physical nature of sludge improves storage characteristics and makes the material far more amenable for disposal to landfill, land reclamation or agriculture. Composting may possibly be economically viable in some circumstances solely due to these factors, however, from an economic appraisal undertaken during the trials described in this paper (Riddell-Black, unpublished data), the process is costly both in terms of capital and operating costs requiring the development and purchase of specialised, automated and highly efficient plant in addition to conventional dewatering facilities. A sewage treatment works producing 10,000 t/year will probably require an investment of about £1 million in order to achieve a reasonably short pay-back period. In order to be cost-effective, revenue will generally be required from the sale of compost to farmers, for land reclamation or amenity purposes, although the sale price is unlikely to be high enough to make composting economically viable without the operational savings that composting offers over other sludge treatment and disposal options. Thus, it is crucial that decisions to develop compost plants must be based on a thorough appraisal of current sludge strategy, particularly costs and ease of disposal, sludge quality, suitable composting systems and potential compost markets.

ACKNOWLEDGEMENT

The work described in this paper was carried out under contract to Thames Water plc, the collaboration and funding for which is gratefully acknowledged.

REFERENCES

(1) de Bertoldi, M., Vallini, G., Pera, A. and Zucconi, F. (1982). Comparison of three windrow compost systems. Biocycle 23 (2) : 45-48.

(2) Finstein, M.S., Miller, F.C., Strom, P.F., Macgregor, S.T. and Psarianos, K.M. (1983). Composting ecosystem management for waste treatment. Bio/Technology 1: 347-353.

(3) Mousty, P., Reneaume, M., and Caille, D. (1985). Stabilization of sewage sludge using various composting processes. Waste Man. Res. 2 : 339-345.

(4) Finstein, M.S. and Miller, F.C. (1985). Principles of composting leading to maximization of decomposition rate, odour control, and cost effectiveness. In: Composting of Agricultural and Other Wastes, J.K.R. Gasser Ed., Elsevier Applied Science, London. 13-26.

(5) Finstein, M.S., Miller, F.C., MacGregor, S.T. and Psarianos, K.M. (1985). The Rutgers Strategy for Composting : Process Design and Control. EPA/600/2-85/059. Accession no. PB85-207538. National Technical Information Service.

(6) Sikora, L.J. and Sowers, M.A. (1985). Effect of temperature control on the composting process. J. Environ. Qual. 14 (3) : 434-439.

(7) Stentiford, E.I. (1987). Recent developments in composting. In: Compost Production, Quality and Use. M. de Bertoldi, M.P. Ferranti, P. L'Hermite and F. Zucconi Eds. Elsevier Applied Science, London. 276-295.

(8) Border, D., Coombes, C. and Shellens, M. (1988). Composting straw with untreated liquid sludge. Biocycle 29 (6) : 54-59.

(9) McKinley, V.L. and Vestal, J.R. (1985). Microbial activity in composting. Biocycle 26 (7) : 47-50.

(10) de Bertoldi, M., Coppola, S., and Spinosa, L. (1983). Health implications in sewage sludge composting. In: Disinfection of Sewage Sludge: Technical, Economic and Microbiological Aspects. A.M. Bruce, A.H. Havelaar, and P. L'Hermite Eds., Reidel, Dordrecht. 165-178.

(11) de Bertoldi, M., Frassinetti, S., Bianchin, L. and Pera, A. (1985). Sludge hygienisation with different compost systems. In: Inactivation of Micro-organisms in Sewage Sludge by Stabilisation Processes. D. Strauch, A.H. Havelaar and P. L'Hermite Eds., Elsevier Applied Science, London. 64-76.

(12) de Bertoldi, M., Vallini, G. and Pera, A. (1983). The biology of composting: a review. Waste Man. Res. 1 : 157-176.

(13) de Bertoldi, M., Vallini, G and Pera, A. (1985). Technological aspects of composting including modelling and microbiology. In: Composting of Agricultural and Other Wastes. J.K.R. Gasser Ed., Elsevier Applied Science, London, 27-41.

(14) Pereira-Neto, J.T., Stentiford, E.I. and Smith, D.V. (1985). Survival of faecal indicator micro-organisms in refuse/sludge composting using the aerated static pile. Waste Man. Res. 4 : 397-406.

(15) Pereira-Neto, J.T., Stentiford, E.I. and Mara, D.D. (1987). Comparative survival of pathogenic indicators in windrow and static pile. In: Compost: Production, Quality and Use. M. de Bertoldi, M.P. Ferranti, P. L'Hermite and F. Zucconi Eds. Elsevier Applied Science, London. 276-295.

(16) Witter, E. and Lopez-Real, J. (1988). Nitrogen losses during the composting of sewage sludge, and the effectiveness of clay soil, zeolite, and compost in adsorbing the volatilized ammonia. Biological Wastes 23 : 279-294.

(17) Bishop, P.L. and Godfrey, C. (1983). Nitrogen transformations during sludge composting. Biocycle 24 (4) : 34-39.

(18) Brady, N.C. (1974). The Nature and Properties of Soils, 8th Edition. Macmillan, New York.

(19) Department of the Environment. (1989). Code of Practice for Agricultural Use of Sewage Sludge.

Table 1 Mass balance for sludge-straw feedstock composted by aerated static piles

Component	Time (days)	Fresh weight (t)	Dry weight (t)	Accumulated total weight loss (t)	Accumulated total weight loss (%)	Accumulated volatile solids loss (t)	Accumulated volatile solids loss (%)
Straw		10.7	8.7				
Sludge		145.1	30.8				
Total material loaded on manifold	0	155.8	39.5				
Material stripped off manifold	24	102.1	31.3	53.7	34.4	8.2	20.7
Second remix	53	71.9	29.3	83.9	53.8	10.2	25.8

Table 2 Volumetric losses during composting of sludge-straw feedstock by aerated static pile

Component	Fresh material (m^3)	Dry material (m^3)	Total accumulated volume reduction (%)	Reduction in volume of dry material (%)	Bulk density
Material loaded on manifold	274.6	158.3			0.14
Material stripped off manifold	140.9	70.1	48.7	55.7	0.22
Second remix	113.9	71.3	58.5	-	0.26

Table 3 Relative heavy metal concentrations of parent sludge, sludge-straw feedstocks and mature compost

	Metal	Sludge concn. (mg kg^{-1})	Feedstock concn. (mg kg^{-1})	Mature compost concn. (mg kg^{-1})	Decrease in feedstock concn. relative to sludge (%)	Increase in compost concn. relative to sludge (%)	Increase in compost concn. relative to feedstock (%)
Static pile	Cu	948	737[1]	1140[1]	22	20	55
	Zn	943	734	1150	22	22	57
	Cd	9.82	7.63	13.3	22	35	74
Turned windrow	Cu	948	772[2]	1198[2]	19	26	55
	Zn	943	787	1349	17	43	71
	Cd	-	5.33	8.22	-	-	54

1 Actual concentration data.

2 Determined from models describing the relationship between volatile solids content and metal concentration using values for VS of 70% for the feedstock and 55% for compost (see Figure 8)

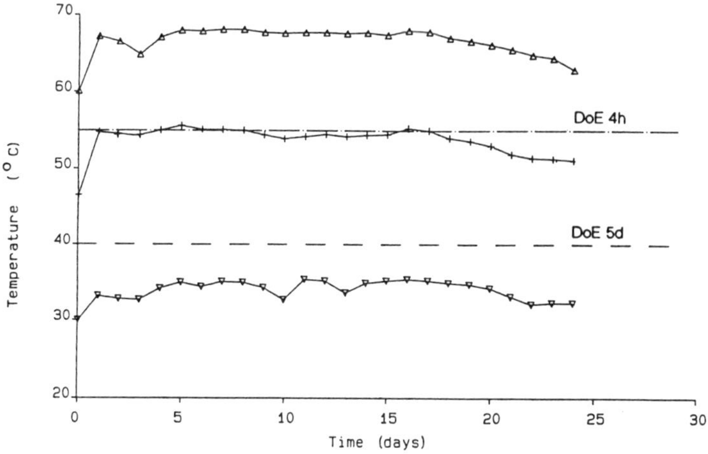

Figure 1 Daily temperature profile of aerated static pile sludge–straw compost
(△ Max, ▽ Min, + Mean)

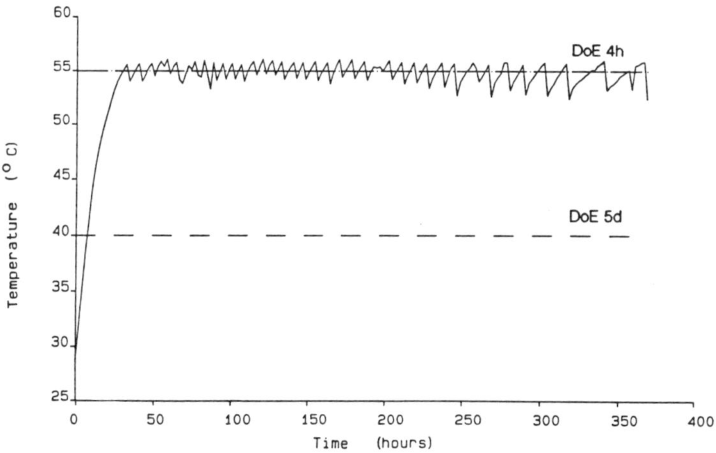

Figure 2 Hourly temperature profile of aerated static pile sludge–straw compost

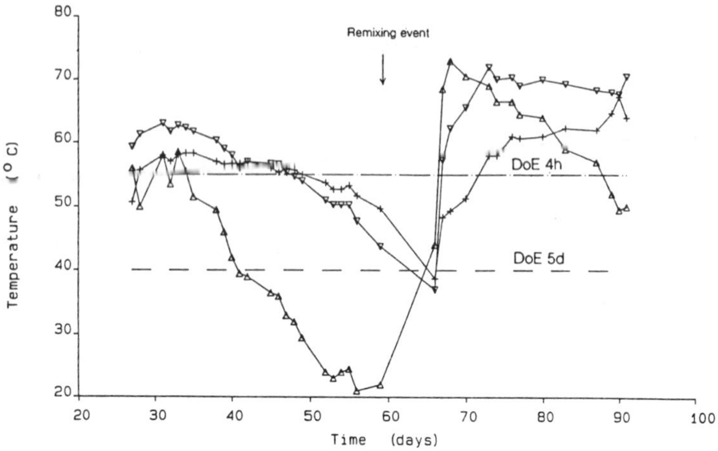

Figure 3 Daily temperature profile of static pile sludge–straw compost after stripping off the manifold (△ 0.5m, ▽ 1.0m and + 1.5m)

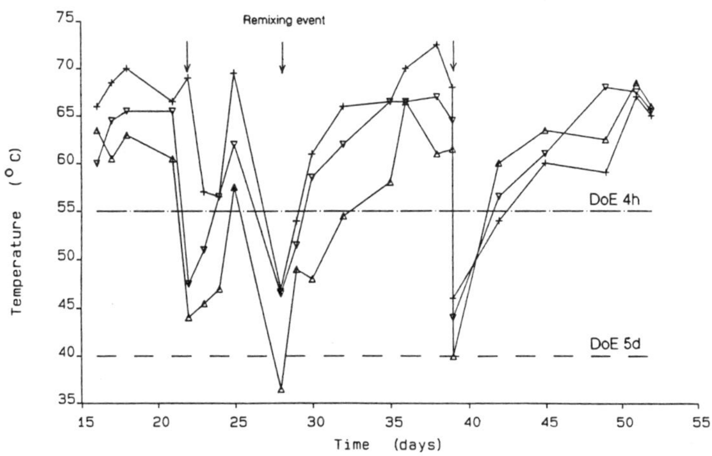

Figure 4 Daily temperature profile of windrowed sludge–straw compost (△ 5%, ▽ 7% and + 8.5% straw)

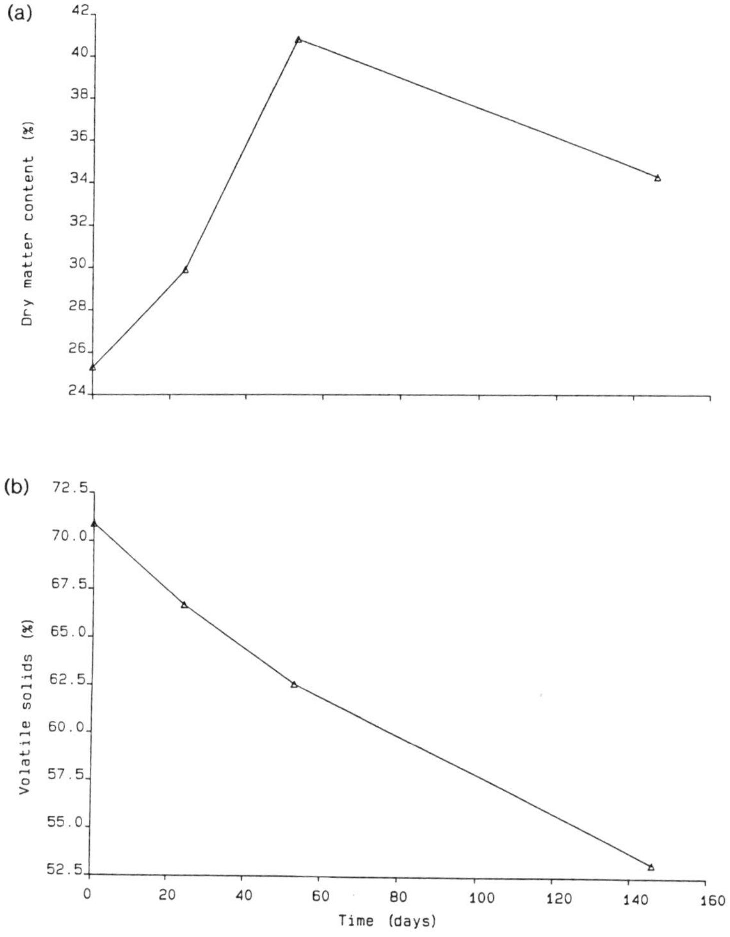

Figure 5 Dry matter (a) and volatile solids (b) content of sludge–straw feedstock composted by aerated static pile

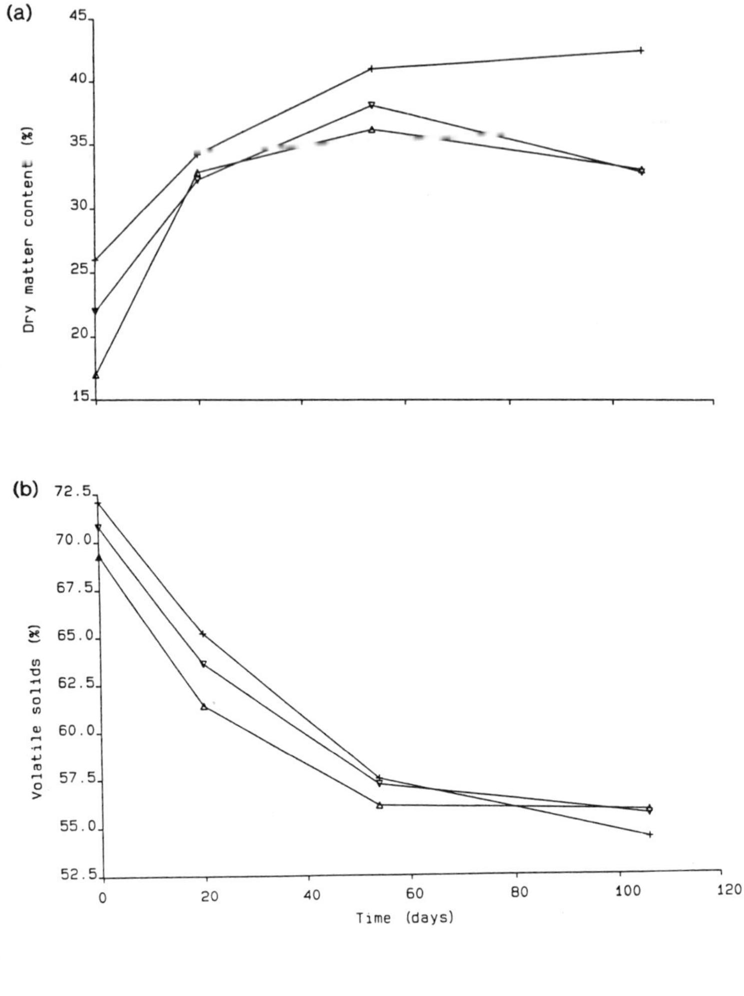

Figure 6 Dry matter (a) and volatile solids (b) content of sludge–straw feedstock composted by turned windrow (△ 5%, ▽ 7% and + 8.5% straw)

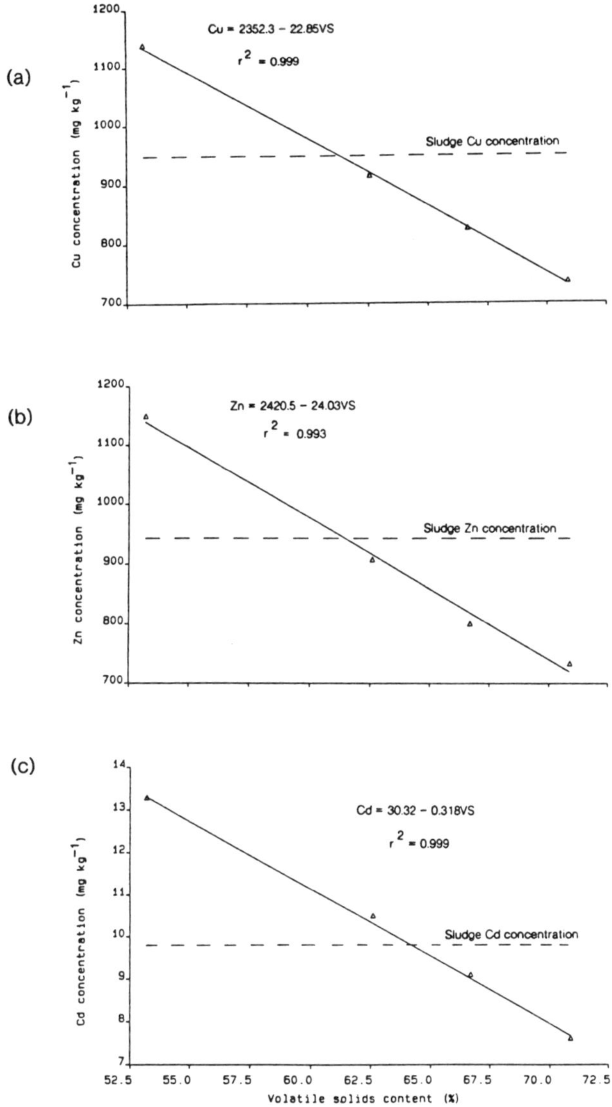

Figure 7 Relationship between the volatile solids content and Cu (a), Zn (b) and Cd (c) concentration of aerated static pile sludge–straw compost

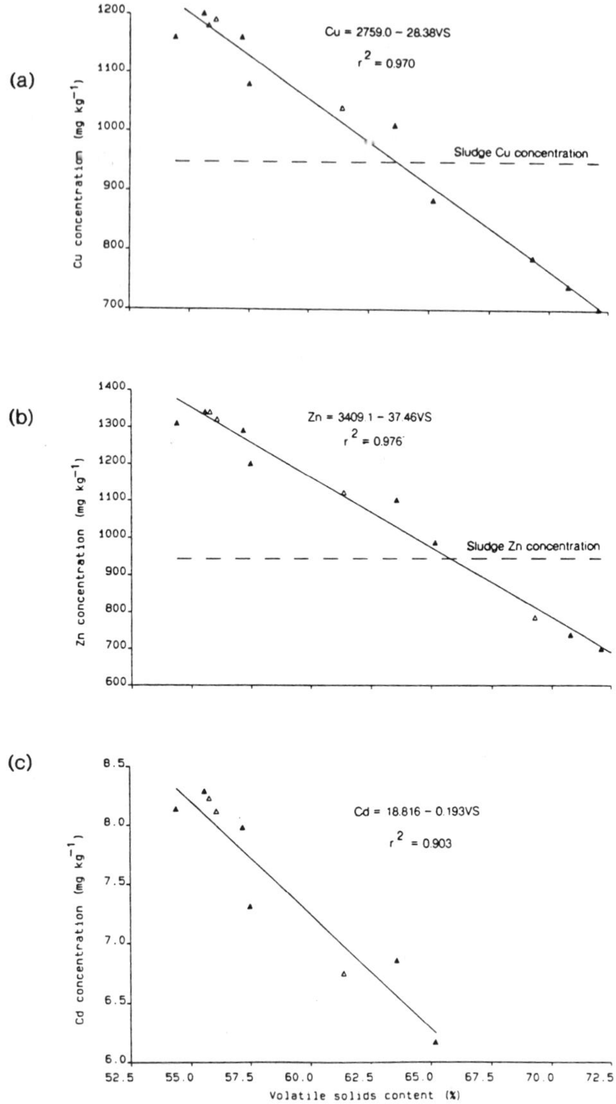

Figure 8 Relationship between the volatile solids content and Cu (a), Zn (b) and Cd (c) concentration of turned windrow sludge–straw compost (△ 5%, ▲ 7% and ▲ 8.5% straw)

INTEGRATED STABILIZATION PROCESSES OF SEWAGE SLUDGES AND URBAN SOLID REFUSES BY CO-LANDFILLING AND CO-DIGESTION

G. BOARI, I.M. MANCINI and S. MASI

Istituto di Idraulica e Costruzioni Idrauliche
Facoltà di Ingegneria, Università della Basilicata
via della Tecnica, 3 - 85100 Potenza (I)

Summary

Sanitary co-landfilling of wastes containing significant amount of biodegradable components, as Municipal Solid Refuses (MSR) and Wastewater Sludges (MWS), requires that operations have to be controlled to optimize the biochemical degradation processes. Researches conducted with simulated landfill test cells or lysimeters have shown that the stabilization rate of the codisposed organic wastes significantly increases with beneficial effects on leachate quality and gas production. The acid phase period can be reduced from many years to few months by co-disposing MSR and MSW, but not suppressed, and the biogas is normally burned without energy recovering. Actually research activities are conducted at Basilicata University for enhancing the acid phase stabilisation of MSR and MWS, co-landfilled in a proper amount, and the biogas and energy recovery, by digesting collected leachates and biological MSW on anaerobic section of municipal wastewater plant. The demonstrative project, financed by Environmental Authority of Italian Government, is based on integrated utilization of different plants, in operation or under rebuilding, to treat the refuses of Potenza town and its sorrounding (Southern Italy). Modalities of co-landfilling operations are tested on pilot lysimeters.

1. INTRODUCTION

Civil works produce two different discharges with high organic content: Municipal Solid Refuses (MSR) and Wastewater Sludges (MWS) from physical and biological treatment. The integrated utilisation of different processes, as inceneration, to burn the MSR fraction at higher heat value, co-landfilling of organic MSR fraction and primary MSW and anaerobic co-digestion, at municipal wastewater treatment plant, of acid leachate and biological MSW, permits a better energy recovery and environmental impact control.

Researches conducted on MSR and MWS co-landfilling and

co-digestion are initially examined. Further a demonstrative project, financed by Environmental Authority of Italian Government, is presented. It is characterized by an integrated utilisation of different plants, in operation or under rebuilding, to treat the refuses of Potenza municipality. Modalities of landfill operation are tested on pilot lysimeters.

2. CO-LANDFILLING

A better knowledge of anaerobic stabilization processes of MSR and their kinetics, permits now to recognize two subsequent and well characterized phases:

I) The acetogenic one at high hydrolysis and refuse stabilisation rates.

II) The methanogenic, with biogas production and leachate self depuration, after the easily soluble part of organic matter has been converted.

The landfills environmental impact, when they are completely filled, can be minimized by optimizing the biochemical processes control.

This can be obtained by increasing refuses water content, also with leachate recirculation, by adding buffers and seeding methanogenic bacteria (1), in order to get as soon as possible a steady methanogenic phase.

Refuses and sludges co-disposal has a beneficial effect, according to many experimental results (2, 3), due to the high water content of the sludges, to theirs buffering and seeding capacity.

On Fig.1 and Fig.2 COD and specific methane production, releaved by Craft and Blakey (3), are reported as time function, for shredded refuses alone and co-disposed with raw dewatered or liquid digested sludges. It can be seen that co-disposal techniques reduces the time to get a stable methanogenic phase, expecially with dewatered sludges, but liquid digested ones promote the methane production and enhance the stabilization rate during acethogenic phase.

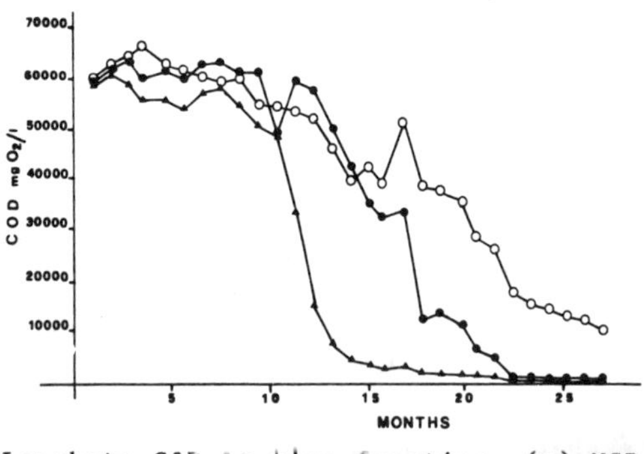

Fig.1 - Leachate COD as time function (o) MSR only; (▲) MSR + raw primary MWS; (•) MSR + liquid digested MWS.

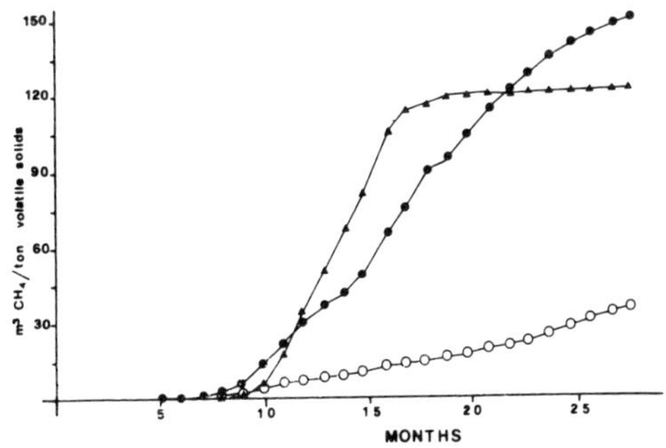

Fig.2 - Cummulative methane production rate (key as in Fig.1).

An integrated approach of MSR and MWS co-disposal, when incineration and co-landfilling processes can be utilized, needs to screen refuses in order to reduce energy consumption and land requirements.

On Tab.I compounds contents and characteristics of raw MSR, on Potenza Municipality, and of the two fractions selected through an 80 mm sieve are reported.

Tab.I - MSR raw and screened fractions through 80 mm sieve: compounds content and characteristics.

	raw	screened	residue
		percent by weight	
water	48.9	58.2	33.6
volatile (VS)	29.0	18.2	49.4
inerts	22.1	23.6	17.0
total weight (tonns/d)	88	55	33
total volume (m^3/d)	159	78	81
COD = 1.4*VS (tonns/d)	36.4	14	22.4
net heating value (kcal/kg)	1021	414	2032

The COD load of leachates, from co-landfilling screened MSR fraction and raw primary sludges, can be estimated assuming a solubilization factor of 25% for the total COD during the acid phase, as reported in Fig.3.

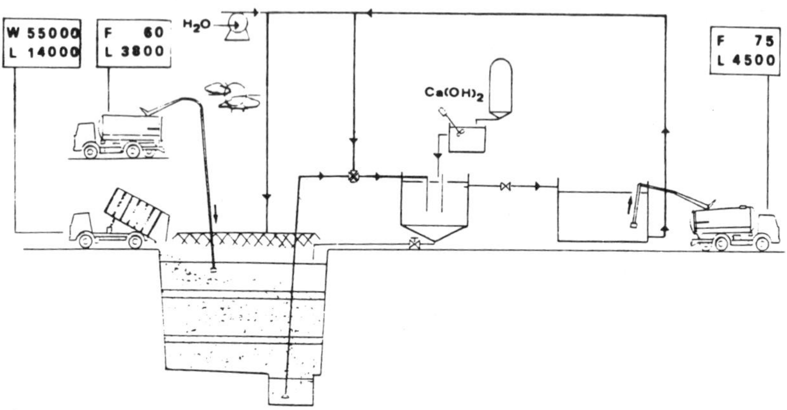

Fig.3 - Flow sheet and COD balance of leachatee on sanitary landfill. F=flow rate (m^3/d);L=organic load (kgCOD/d); W=screened refuse disposal (kg/d).

3. CO-DIGESTION OF LEACHATES AND SEWAGE SLUDGES

Studies conducted on co-digestion of olive oil mill effluents (4), or concentrated petrochemical wastes (5) rich in volatile fatty acids, as leachates from acetogenic phase, have shown that volumetric COD loading of sludges anaerobic digesters, can be increased 1.5-2 times with a removal of soluble COD as high as 85%.

Anaerobic treatment of leachates, on different reactors (conventional, UASB, anaerobic filters, etc.) has been tested with COD removal efficencies as above reported for the co-digestion process (6, 7). On Tab.II we have reported the project parameters that can be assumed now and in the next future for some of the anaerobic digesters.

Leachate disposal at municipal wastewaters treatment plant permits also to utilize aerobic biological reactors for the removal of residual BOD_5 and ammonia-nitrogen.

In Fig.4 hydraulic and organic loads are reported, referring to a wastewater treatment plant serving 100,000 inhabitants, as for Potenza municipality, with a pro-capita COD of 125 g/p*d, for estimating a mass and energy balance of a conventional process.

Tab.II - Project parameters at the present and in the future for some anaerobic digester of leachates, compared with these of sludge digester (digestion temperature 35°C).

Anaerobic digestion process	Project parameters					
	present			future		
	C	B	M	C	B	M
Anaerobic contact for surplus sludges	2.9	0.6	0.30	6.0	1.3	0.30
Anaerobic contact for leachates	≤4	1.6	0.30	(*)	(*)	(*)
Leachates and surplus sludges co-digestion	4.4	1.1	0.30	9.0	2.5	0.30
Fluidized bed (UASB) for leachates	11.4	3.5	0.30	15.0	5.0	0.30

C = volumetric organic load ($kgCOD/m^3r*d$)
B = specific biogas production (Sm^3/m^3r*d)
M = specific methane production for kg CODr ($Sm^3/kgCOD$)
(*) load increases are not possible due to inhibition problems

The same balances were done, as shown on Fig.5, for co-digesting biological MWS and acid leachates derived from co-landfilling of screened MSR and raw primary MSW. As Tab.I shows screened MSR are more rich in COD and water content than raw one.

Specific energy consumption of activated sludge process was fixed like 1.1 kWh/kg COD removed taking into account the simultaneous nitrification process. Energy recovered by biogas was estimated assuming a 40% COD removal on the digestion of biological MSW and 85% of leachates; the methane yield was fixed as 0.30 Sm^3/kg COD removed and the overall efficiency for energy production as 0.26.

Volume increase of biological reactors is limited, due to the high specific operative loads of anaerobic co-digester, to the co-landfilling of raw primary MSW on landfill and to the low residual BOD_5 of leachate to be removed aerobically.

Settling tanks are not affected as well due to the small daily leachate volumes. It is remarkable to note that electric energy produced by biogas, when leachates are co-digested, covers the 50 % of that consumed by the aerobic process for urban wastewaters treatment. Energy recovery, without leachate co-digestion, is about 30%.

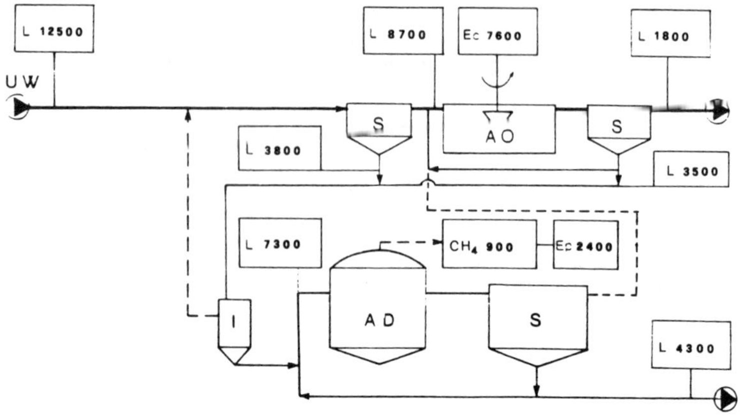

Fig.4 - Daily organic substrate L (kg COD), energy consumption E_c (kWh) and energy production E_p (kWh) for conventional wastewater process.

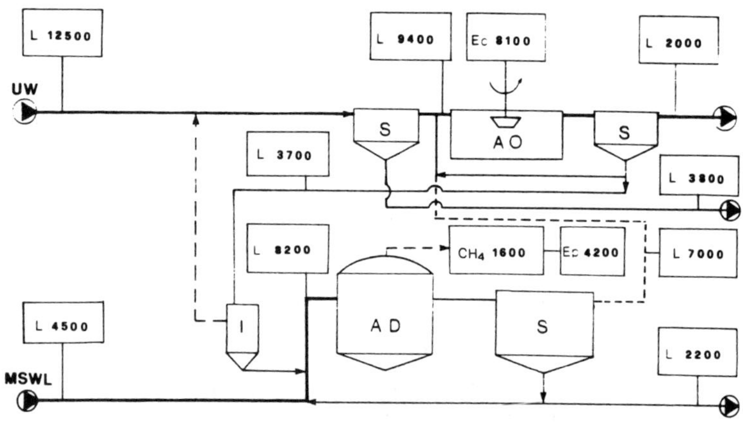

Fig.5 - Daily organic substrate L (kg COD), energy consumption E_c (KWh) and energy production E_p (kWh) for conventional wastewater process and co-digestion of biological MWS with acid leachates.

4. DEMONSTRATIVE PROJECT

The demonstrative project, promoted by Basilicata University and financed by Environmental Authority of Italian Government, is based on an integrated utilization of three facilities, in operation or under rebuilding, to treat refuses from Potenza municipality.

These facilities are:

- Two incinerators of 60 tonns/d each, under rebuilding, to facilities for post-combustion at 950 °C, filtration of particulate emission and adsorption of organic compounds. It was estimated that post-combustion process of raw MSR from Potenza municipality, will require 38 kg of gas oil. On the contrary, incineration of the residue from 80 mm sieve will allow an energy recovery equivalent to 30 kg of gas oil. Furthermore it permits to reduce the landfill volume of 50%.
- A lined sanitary landfill, divided in two section of 60,000 m^3 and 40,000 m^3 respectively, is now in operation and receives in the first section raw MRS, due to incinerators rebuilding. The second section, initially projected to receive incinerators ashes, will be utilized to co-dispose screened MSR and primary MSW.
  Researches are conducted on three lisymeters, filled with about 5 tonns of co-disposed refuses, described in other paper (8). Their aim is to enhance acid phase stabilisation of screened MSR and primary MWS, co-landfilled in a proper amount, and to obtain the operating parameters for optimizing stabilization processes of the demonstrative landfill.
- The anaerobic section of municipal wastewater treatment plant where leachates, transported with the same tank truck utilized to dispose primary MSW on the landfill, will be digested utilizing for research purposes the very old Imhoff digesters, in operation since 1934 as second full scale digester in Europe.

REFERENCES

(1) BUIVID, M.G., WISE, D.L., BLANCHER, M.D., REMEDIOS, E.C., JENKINS, B.M., BOYD, W.G. and PACEY, J.G. (1981). Fuel gas enhancement by controlled landfilling of municipal wastes. Resources and Conservation, 6, 3-20.

(2) FARREL, J.B., DOTSON, G.K., STAMM, J.W. and WALSH, J.J. (1987). The effects of municipal wastewater sludge on leachate and gas production from sludge-refuse landfills. Personal communication to Sludge Network Members.

(3) CRAFT, D.G. and BLAKEY, N.C. (1988). Codisposal of sewage sludge and domestic waste in landfills. ISWA Proceedings, Copenhagen.

(4) CARRIERI, C., BALICE, V., BOARI,G., ROZZI, M. and SANTORI,M. (1986). Anaerobic treatment of olive mill effluents and sewage sludge in conventional digester. Proc. of the International Symposium on Olive By-products Valorization, Sevilla.

(5) DE BAERE, C., VAES, H. and VERSTRAETE, W. (1982). Combined digestion of concentrate industrial wastes and municipal wastewater sludges. Proc. of the International Symposium on Advances in Anaerobic Digestion, Mexico City.

(6) JEANS, J.M., Van de SCHROEFF, A. AND JAAP,A. (1987). A treatment concept for leachate from sanitary landfills. ISWA Proceedings, Cagliari.

(7) BLAKEY, N.C. and MARIS, P.J. (1987). On-site leachate management - Anaerobic processes. ISWA Proceeding, Cagliari.

(8) BOARI, G. and MANCINI, I.M. (1987). Leachate stabilization and disposal at sewage treatment plants. ISWA Proceedings, Cagliari.

INTEGRATED TREATMENT OF SOLID WASTES, SEWAGE SLUDGES AND LIQUID MANURE

J.J. van den Berg
H.W. de Mooij
Grontmij n.v. consulting engineers
De Bilt, Netherlands

SUMMARY

In the Netherlands new techniques for the treatment of wastes - in particular sewage sludge - are being developed such as incineration and the wet oxidation. For the time being alternatives have to be used for the disposal of sewage sludge.
Not only sewage sludge but also liquid manure and municipal wastes are present in large quantities. A project is in preparation for the combined treatment of municipal wastes, sewage sludge and liquid manure.

1. INTRODUCTION

In the Netherlands the disposal of waste is gradually becoming an almost insurmountable problem.

Due to the present economic boom the amount of municipal and industrial waste and other waste is growing fast. However, due to the increasing environmental awareness, a growing number of restrictions is imposed upon available methods of waste disposal, e.g.:

- new waste disposal tips have to be fully isolated from the surroundings and leachate has to be treated before discharge on surface water;
- flue gas from municipal waste incinerators has to be treated to a large extent. Present Dutch emission standards are among the strictest in the world;
- compost from municipal waste or municipal sewage sludge is not to be used for agricultural purposes any more, due to its high content of heavy metals;
- due to the increasing concentration of nitrates in ground water (threatening two thirds of the drinking water supply in the Netherlands), disposal of waste from the livestock industry as fertilizer in agricultural areas must be reduced by 6 million tons per year by 1994.

Although economic implications of said measures are not fully neglected, it is accepted in the Netherlands that strong measures are requested now to avoid damage to the environment that in future will turn out to be irreparable. This not only affects the cost of waste disposal, but in some instances even has a much wider impact. For instance in the last three months three municipal waste incinerators have been closed down as the emission of dioxins from these installations were considered to be unacceptable for the environment. Furthermore the Ministry of Housing, Physical Planning and Environment has announced an imposed restriction on livestock farming in 1994 if the goal of reduction of wastes from this branch has not been reached by that time.

In reaction to the present attitude towards the environment, many innovative projects are initiated to respond to the changing market of environmental technology. Techniques that only years ago were considered to be too expensive for full scale use, now are within the scope of feasibility. Also the overall attitude towards waste handling is gradually changing from waste disposal towards waste management. Where unto recently mere disposal of waste was the most economic action, today a general approach is applied, which includes production (waste prevention), reuse, recycling, energy recovery etc.

2. WASTE HANDLING SITE BOELDERSHOEK

One of the projects presently under design in the Netherlands is the waste handling station Boeldershoek. This project not only draws attention because of the innovative techniques that are used, but also because it is an example of an integrated approach to waste handling in which the combined handling of several types of wastes offers distinct advantages over separate handling.

At Boeldershoek the following types of waste will be processed:

- municipal waste and comparable waste from industries;
- municipal sewage sludge;
- livestock manure.

The installation will eventually consist of the following units:

- Production of Refuse Derived Fuel (R.D.F.)
  The municipal waste and the comparable waste from industries are separated into an R.D.F. fraction, a metal fraction and a rest fraction;

- Energy production
  The R.D.F. is converted to steam, which is subsequently used for the drying of manure and to cover internal energy demand. A surplus of energy production will be supplied to the local electricity net;

- Sludge processing
  The municipal sewage sludge is dried in a rotating drum and subsequently incinerated. The heat that is generated in this way is used for the drying of the sludge;

- Manure processing
  The energy produced by the R.D.F. is used for the drying of the manure. Before thermal drying the manure is dewatered by means of belt presses. After drying the manure pellets are sold as fertilizer;

- Tip site
  All waste products that are unsuitable for reuse or recycling are tipped at the tip site adjoining the location.

- Wastewater treatment plant
  All waste water flows from the installation are treated in the waste water treatment plant. This plant consists of:
  * physico-chemical treatment of condense water from sludge drying;
  * aerobic biological treatment of surplus water from manure processing;
  * micro filtration, reversed osmosis and evaporation of effluent of the aerobic biological treatment;
  * anaerobic biological treatment of acid leachate from the tip site.

  The effluent from the wastewater treatment plant is discharged into the local sewerage system and will be further treated in a communal wastewater treatment plant.

- Composting plant (optionally)
  Separated collection of biodegradable wastes is becoming more and more customary in the Netherlands. At the location Boeldershoek space has been reserved for realisation of a composting plant for this kind of waste in the near future. As compost obtained from integral muncipal waste is regarded unsuitable for agriculture, composting of integral municipal waste is not considered to be a valid option.

A flow diagram for the complete system is given in figure 1. The energy diagram is given in picture 2.

3. SPECIFICATIONS OF THE VARIOUS UNITS

3.1. Production of Refuse Derived Fuel

A total amount of 230.000 tons per year of municipal waste from participating communities and company wastes will be processed at Boeldershoek. The two waste streams plus a stream of gross wastes are stored in three separate bunkers. The bunkers have a storage capacity of 2 days. Municipal and company waste will be processed on two separated lines. After shredding and sieving the two R.D.F. streams are mixed, metal is magnetically removed and the R.D.F. is transported by belt to the fuel bunker of the power unit. The inert material together with the gross waste is transported to the tip site. Approximately 75% of the municipal waste and 95% of the company waste will remain as R.D.F.

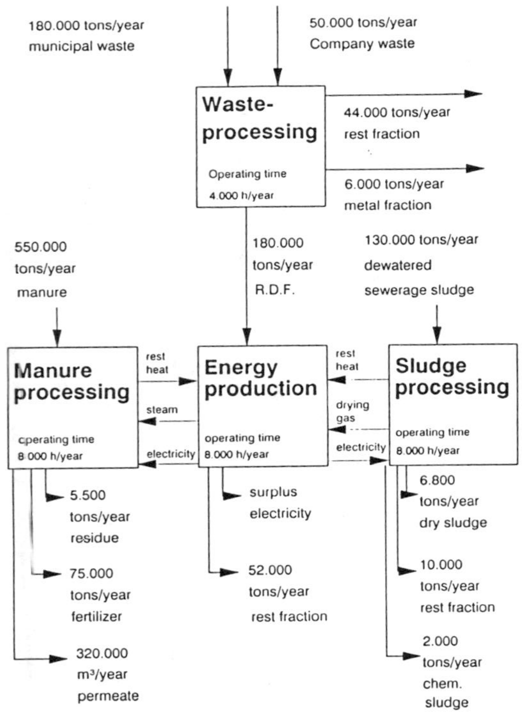

Fig 1 - Mass flow

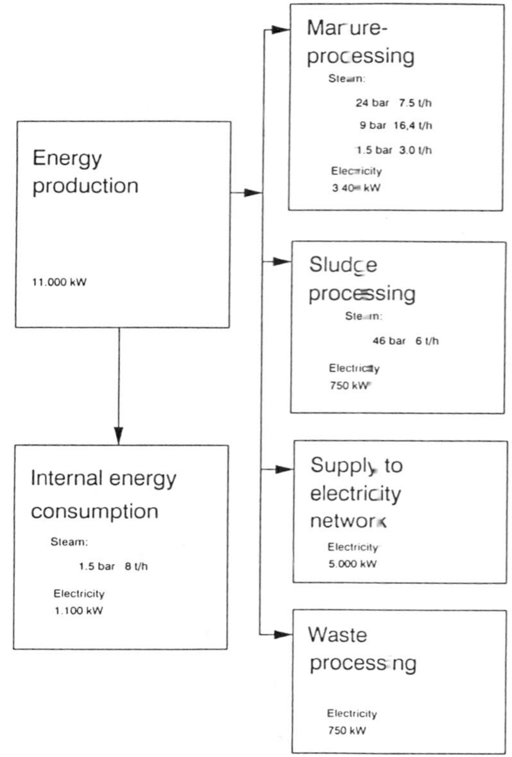

Fig 2 - Energy diagram

3.2. Energy production

The energy from R.D.F. is used to produce steam in two combined incinerators and steam boilers. The total capacity for steam of the system is 70 tons/hour at 40 bar. 64 tons/hour are used to generate electricity (11 MW). The remaining part (6 tons/hour at 40 bar) is used for the drying of sludge and the reheating of treated flue gasses. The worked out steam from the generators is used for the manure processing, sludge drying and other purposes.

The produced electricity is partly used for the internal energy demand (approximately 6 MW). 5 MW can be supplied to the local electricity network.

The incinerators are of the grate type, with a capacity of 12,5 tons/hour each. Ventilation air from the R.D.F. production area and the various bunkers and the drying air from the sludge processing unit is used as primary combustion air to reduce odour nuisance. The incineration temperature lies between 850° and 1.000°C. During start up procedures and stop procedures additional gas burners are used to maintain this temperature throughout the incineration process.

A summary of the energy production unit is given in table 1.

3.3. Flue gas treatment

Special attention is given to the purification of the flue gas from the incinerators. In order to assure emission concentrations in compliance with the present emission standard a completely dry system of purification is used. This system consists of two stages:

1. Mixing calcium hydroxide and activated carbon with the flue gas and subsequent filtering of the flue gas by means of a fabric filter. In order to assure good operation the flue gas has to be cooled down to 150°C before activated carbon can be added. Acid components (HCl, HF, SO_2) will react with the calcium hydroxide, whereas heavy metals and dioxins will adsorb to the activated carbon. As a film of calcium hydroxide and activated carbon will form on the filter area, a superb contact between flue gas and additives will be assured, rendering high removal efficiencies. With the fabric filter fly ash, calcium hydroxide, activated carbon and adsorbed components will be removed from the flue gas. A dust removal efficiency to a concentration below 5 mg/m^3 is feasible. Dioxins must be removed to a concentration below 0,1 ng/m^3. Both incinerators will be equipped with a fabric filter.
2. NO_x will be removed by means of selective catalytic reduction (S.C.R.). In this process ammonia is added to the flue gas, which in presence of a catalytic medium will react with NO_x to nitrogen (N_2) and water. As this reaction requires a temperature of about 320°C, reheating of the flue gas is required.

Table 1. Energy production

Incinerators		
Municipal waste supply	180.000	tons/year
Moisture contents	28,7%	
Ash contents in dry solids	20%	
Ashes	36.000	tons/year
Primary air	140.000	Nm³/h
incl. sludge processing	160.000	Nm³/h
Surplus air	75%	
Flue gas production	180.000	Nm³/h
incl. sludge processing	200.000	Nm³/h
Steam production	70	tons/year
taken from furnace	6	tons/year
Turbine		
Steam supply	64	tons/year
steam effluent		
- steam 24 bar	7,5	tons/year
- steam 10 bar	16,4	tons/year
- steam 1,5 bar	9,8	tons/year
- steam 0,1 bar (to condensor)	30,3	tons/year
Electricity	max. 11	tons/year

3.4. Sludge processing

A process flow diagram is given in figure 3. The sludge processing installation has a capacity of 160.000 tons/year at a dry solids content of 20 to 30%. The sludge is stored in a bunker and from there transported to a rotating drum oven. Before entering the oven, the sludge is mixed with already dried sludge to a solids content of 50%. In the oven the sludge is further dried to 90% dry solids. Next the sludge is transported to a chain grate incinerator. The flue gas is led to the rotating drum oven and is used for the sludge drying. Next the gas is treated in a multi cyclone and cooled down to 65°C in a scrubber, causing condensation of a large part of the water content. Approximately 36% of the gasses (with an oxygen content of 6%) is led back to the incinerator. The remainder is treated together with the flue gas coming from the municipal waste incinerator in the dry flue gas purification system. The water coming from the wet scrubber is treated in a physico-chemical installation. The energy balance of the sludge processing unit is given in figure 4.

3.5. Manure processing

Approximately 500.000 tons of manure with a dry solids content of 10% will be treated yearly. The manure will be dewatered in belt presses to a dry solids content of 35%. The remaining manure is then further treated in a fluidized bed dryer. For this energy from the municipal waste incinerator is used. All emerging water flows are treated in an aerobic biological water treatment installation, where most of the ammonium is converted to nitrates. Surplus sludge is mixed with the incoming manure for combined treatment. The effluent is further treated in a three stage physical treatment plant, consisting of:

1. micro filtration;
   concentrate is led back to the biological treatment, the effluent is further treated.
2. reversed osmosis;
   concentrate is further treated, permeate is flushed into the local sewerage system.
3. Evaporation;
   the concentrate from the reversed osmosis is boiled down to a brine in a three stage evaporator. The destilate can be further used as process water.

3.6. Tip site

All inert waste products are stored at the local tip site. The site is equipped with high density polyethylene (H.D.P.E.) lining and drainage systems to isolate the waste and leachate from the surroundings. After completion of the site, the site will be covered with H.D.P.E.) to avoid further production of leachate. Until completion of the municipal waste incinerator organic municipal waste is also stored at the site. To date the resulting leachate is treated in a physico-chemical treatment

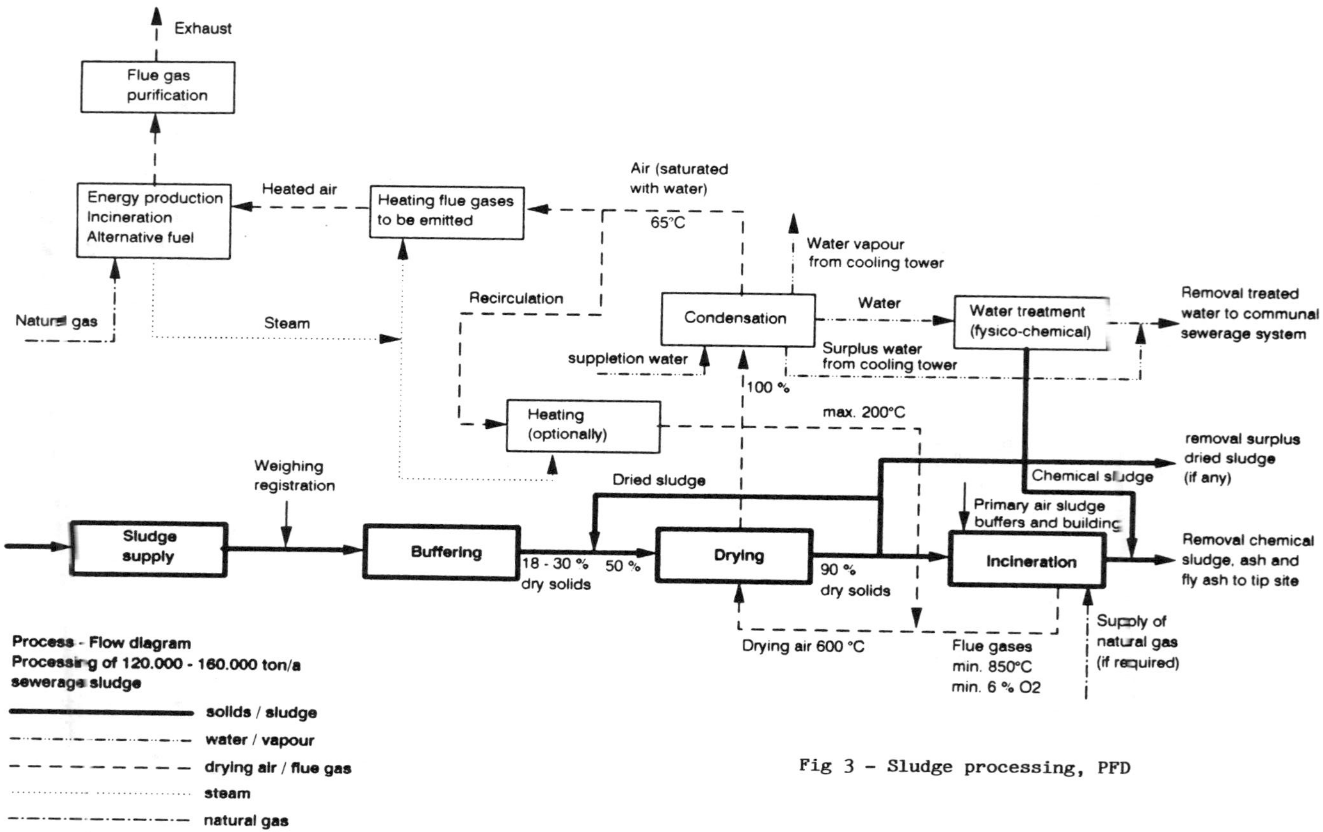

Fig 3 - Sludge processing, PFD

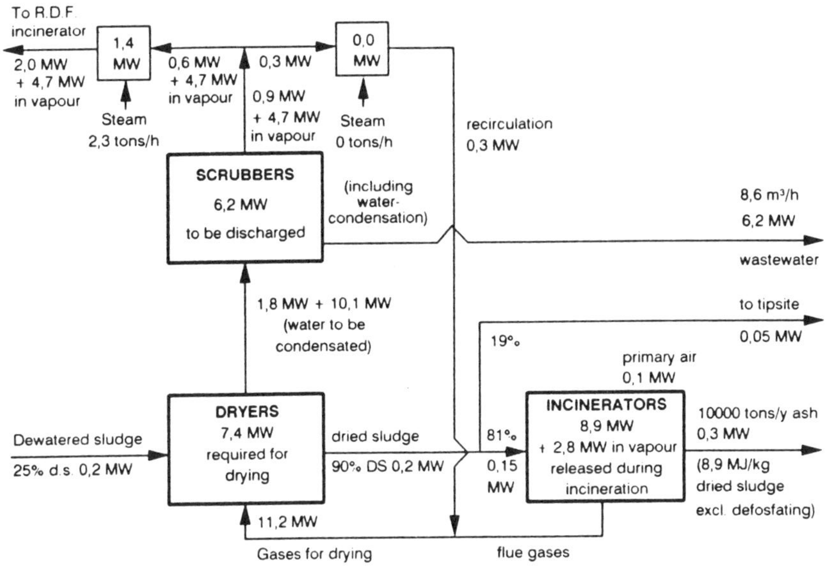

Fig 4 - Sludge processing, energy balance

unit, followed by an upflow anaerobic sludge blanket reactor (U.A.S.B. reactor). In this reactor the chemical oxygen demand is reduced by 70-80%. The effluent is led to the local sewerage system. In future, when no more organic materials are stored, the U.A.S.B. can be taken out of production and treatment will be limited to the physico-chemical unit.

4. ADVANTAGES OF COMBINED PROCESSING OF WASTE STREAMS

At the location Boeldershoek municipal waste, sewerage sludge and manure will be treated simultaneously. The advantages of this combination are manifold:

- Logistical advantages
  Due to the concentration of several processes at one location, transport of waste products can be minimized as the end products or by-products of municipal waste separation, municipal waste incineration, sewerage sludge incineration and manure drying (and in future composting of the organic fraction of municipal waste) can be processed simultaneously at the same location as well.
- Energetic advantages
  The energy obtained by the incineration of municipal waste can be used in the drying processes of manure and sludge. Surplus heat energy production can be converted to electrical energy and fed to the local electricity network.

With the combination of municipal waste incineration and sewerage sludge incineration a flexible system of energy production and use is created.

- Economical advantages
  Several installations can be used for various purposes, for instance the combined treatment of water flows from different sources, combined treatment of flue gasses, combined electricity and steam systems, etc.
- Environmental advantages
  Due to the optimum reuse of heat from the incinerators, less energy is required for the drying of sludge and manure. Also odour nuisance to the surroundings will be greatly eliminated by using all ventilation gasses as primary air for the incinerators.

Literature

1. Grontmij nv.
   Milieu-effectrapportage reststoffenverwerking Boeldershoek (Environmental impact study Boeldershoek), geïntegreerde verwerking van afval mest en slib.
   De Bilt, january 1990.

2. Grontmij nv
   Haalbaarheidsonderzoek mestverwerking Gelderse Vallei (feasibility study manure processing Gelderse Vallei), milieu-aspecten van mestverwerking in de gemeente Ede.
   De Bilt, december 1986.

3. Grontmij nv/RPD.
   Droge reiniging van rookgassen uit afvalverwerking.
   De Bilt, june 1987.

4. v.d. Berg, J.J. and v.d. Schroeff, J.A.,
   Dry cleaning of flue gases in municipal solid waste incinerators.
   Proceedings of the 8th World Clean Air congress, Amsterdam, 1989.

5. D.B.W./R.I.Z.A.
   Verbranding van slib van communale rioolwaterzuiveringsinrichtingen (incineration of sludge from communal wastewater treatment plants).
   Nota nr. 87.060. Lelystad, february 1988.

6. Ministry of housing, physical planning, and environment (NL),
   Richtlijn Verbranden '89 (Guidelines for emission standards 1989).
   The Hague, february 1985.

INTEGRATED SOLID WASTE AND SLUDGE TREATMENT BY WABIO-PROCESS

A. STENRROOS and H. KUTINLAHTI

Outokumpu EcoEnergy Oy

Summary
Biofraction of municipal waste and sewage sludge has been handled in WABIO-process. Waste material is first precomposted and then feeded to biogas reactor. In pilot-scale system with composting time 2 - 4 d and with organic loading of biogas reactor 7 kg VS/M^3 d in 34°C VS-reduction was 72 %. Biogas production was 0,47/kg VS. Input TS of waste material was about 23 % and output TS was 10 %. A full scale plant has been built to Vaasa.

1. INTRODUCTION

The WABIO-process is a two-stage biological method for treatment of municipal waste, sewage sludge and other organic material. In the process waste material is first pre-composted and then fed to biogas reactor. Biogas reactor is so called dry fermentor where input total solid (TS) concentration is 20 - 30 % (w/w).

The WABIO-process has been tested with pilot-scale studies in years 1986 - 1987 with 1 m^3 composter and 5 m^3 biogas reactor. After these studies a full scale WABIO-plant has been build to Vaasa town to handle 12.000 t/a biofraction of municipal waste and 15.000 t/a dried sewage sludge.

2. PILOT-SCALE STUDIES

In pilot-scale studies biofraction of municipal waste and sevage sludge was feeded to WABIO-process. The pilot apparatures were :

- composter 1 m^3
- mass pump 2 m^3/h
- biogas reactor 5 m^3

Filling degree in composter and biogas reactor was 50-60 %.

Input waste was as follows :

- waste paper etc 23 kg/d
- foodwaste etc 12 kg/d
- sludge raw (4 % TS) 55 kg/d

Retention time in composter was 2 - 4 d and temperature was 33°C. Organic loading in biogas reactor was 7 kg VS/m^3 d and temperature was 34°C.

Total VS-reduction in pilot-scale WABIO-process was 72 % and biogas production from input waste was 0,47 m^3/kg VS.

Methane content of the biogas was 58 %. Input TS-content to biogas reactor was 20 - 25 % and output was about 10 %.

3. CONCLUSIONS

The WABIO–process is very efficient method for handling biofraction of municipal waste as sewage sludge.

In composting unit waste materials is homogenized and heated. There also happens sorting of organic compounds. VS_reduction in this unit is very low. In biogas reactor happens stabilization of waste and organic material is converted to methane and carbondioxide.

Stabilization degree of waste after WABIO–process is very good because of two–stage treatment and long retention time in biogas reactor. Retention time is long because input TS of waste material is high compered to conventional biogas reactors.

A full scale WABIO–plant has been built near Vaasa town to handle municipal waste and sewage sludge. Compered to pilot–scale system biofilter has been added to handle output air from composter and also a screw presser has been added to dry the outcoming mass. About 12.000 tn/a biofraction of municipal waste and 15.000 tn/a dried sewage sludge is to be handled in two lines. Pretreatment plant and composting part and biofilter has been tested and in may 1990 waste feeding to biogas reactor has been started.

SESSION II

GENERAL DISCUSSION

REGARDING PAPER 5

L SPINOSA

What effect does composting have on the heavy metal content of the finished product?

J E HALL

As I mentioned in the paper there is a significant bulk reduction during composting as a result of moisture and volatile solids loss. This causes a corresponding increase in heavy metal content of the finished product.

J J VAN DEN BERG

Have you carried out a mass balance on the straw and sewage sludge content during the composting process? Also is it necessary to turn the material during composting to keep the temperature to below about 55°C.

J E HALL

It is a very difficult exercise to carry out a mass balance on the straw and sewage sludge during composting. This is because of the differing volume to weight ratios of the two materials. As regards temperature control during composting, it is not normally necessary to turn the pile during composting to achieve this when you are using forced aeration. In this case the aeration prevents the temperature rising too high.

Y J PUOLANNE

Have you considered the costs of operating the various options of composting with the quality of the finished product?

J E HALL

In general the only quality criterion considered is the hygenic effect and that is dependent on the temperature reached during composting.

J J VAN DEN BERG

At present farmers receive sewage sludge free but have to pay a small amount for composted material. Do you see a time when farmers would be prepared to pay reasonable sums for both sewage sludge and composted material?

J E HALL

That could happen in the future but not until farmers are confident, both in terms of the continuation of supplies and of the quality (ie lack of risk), of these materials.

P OCKIER

Is waste refuse composting practised in the UK and does the heavy metal content restrict the markets/outlets available for the finished product?

J E HALL

Only small amounts of waste refuse are composted in the UK. The heavy metal content of the composted material may certainly restrict the outlets available. However, similar constraints which apply to the disposal of sewage sludge would also limit the disposal of composted waste refuse.

L LANGGAARD

The supply of straw in Denmark is limited as it is used as a fuel in power plants. Does a similar situation exist in the UK?

J E HALL

At present straw is freely available in the UK. There is no real interest in it as an energy source.

L SPINOSA

What is the C:N ratio of the composted material?

J E HALL

This is not really optimised for plant nutrition as it is about 12:1. Attempts to improve the situation can mean that the composted material becomes too dry and moisture limiting when used as a fertiliser.

REGARDING PAPER 6

J E HALL

How is the sludge and waste mixed in the lysimeter? Also is it mixed homogeneously?

G BOARI

The sludge and waste is mixed by hand in the ratio 1:4. It is not mixed homogeneously but put in layers. The water content of the mixture is very important for effective digestion. A large reduction in volume of the solids occurs and that is a very useful feature of the process.

F COLIN

Are there any odour problems with the operation of your plant?

G BOARI

No, we are very pleased about the lack of odour caused by the plant.

H CAMPBELL

I am slightly surprised that the biogas produced in your process is burnt without any attempt of energy recovery. Is this because the amount of biogas produced is insufficient to make energy-recovery worth while?

G BOARI

Burning the biogas produced without energy recovery only exists for the pilot plant. I am sure that energy recovery from the biogas would be incorporated in a full scale plant.

REGARDING PAPER 7

P GEUZENS

Have you considered the different energy requirements for the various wastes treated?

J J VAN DEN BERG

Yes, these have been considered. We have also undertaken an environmental impact assessment for the process.

H CAMPBELL

In the calculations of the costs of your process have you included the revenue from the sales of treated product?

J J VAN DEN BERG

No, these have not been included.

J E HALL

Do you consider the production of Refuse Derived Fuel (RDF) as an important source of energy?

J J VAN DEN BERG

Yes, very important. It can mean that the plant effectively produces a surplus of electrical energy.

REGARDING PAPER 8

G BOARI

Is it possible to pump the material in your process as the solids content is as high as 22%? Also is there a substantial reduction in the volume of the solid waste and sludge mixture during the process?

A STENROOS

In fact, the solids content in the process can be as high as 30% and even at that level it is still possible to pump the material. The extent of the volume reduction during the process depends very much on the quality of solid waste used but can be up to 20-30% of the initial volume.

H CAMPBELL

In your process, you first compost the material and then treat it anaerobically. Surely this would mean the loss of some of the carbon content as carbon dioxide? Would it not be preferable to treat the mixture anaerobically first?

A STENROOS

The order of the processes is dictated by the oxygen content of the mixture. It is necessary to compost the mixture first when the oxygen content is high so that the cellulose content present will be hydrolysed to the greatest extent.

OVERALL SUMMARY AND CONCLUSIONS

A.H. DIRKZWAGER

Institute for Inland Water Management
and Waste Water Treatment
Lelystad, The Netherlands

and

P. OCKIER

Flemish Society for Water Treatment
Oostende, Belgium

Due to the limited time at the Workshop the programme only covered a selection of topics on the subject "Recent developments in sludge processing". For an exhaustive inquiry, a conference of several days would be required.

The first section of the session concerned the following themes : assessment of the pollution potential of dumped sludge, defining a standard for the concept of congealed sludge, optimizing sludge dewatering by a PC–based operator system and the effect of stabilization processes on sludge dewatering and hygienization.

The second section of the session was applied to practical situations in the field of WORKING PARTY 1, looking for systems, based on co–treatment and co–disposal methods for optimizing techniques and reducing costs. It indicates a disposition to look more than in the past for special co–treatment and co–disposal facilities for sewage sludge and other waste substances, if the direct use of sewage sludge and, for instance, animal manure will no longer be possible, due to more stringent standards. These developments, adapted to specific circumstances, conditions and local situations generally call for accurate preparation and study of special problems. In order to make a choice from the various possible techniques, their efficiency, cost, environmental impact, and completion time all need to be taken into account.

The paper of F. Colin, who was decorated for his research activities in France, describes a possible approach for the study and forecasting of the liberation of dissolved pollution by landfilled waste sludges. The three levels of approach, i.e. laboratory tests, tests in pilot columns, and use of large boxes or lysimeters provide complementary information in the given order. Laboratory tests are chiefly able to assess a pollution potential of the landfilled sludges, and are not suited for studying long–term behaviour. To forecast the liberation kinetics of this pollution, at least experiments in pilot columns have to be executed.

Very interesting are the numerous works on studying the co–landfilling of sludges and household refuse and the study of the influence of the nature and disposition of an additional material dumped with sludges. Colin found, that in a certain number of cases and for certain pollutants, sewage sludge added to domestic refuse improves the results, in comparison with the dumping of refuse alone. The quantities of solubilized pollutants are lower.

The paper of P. Geuzens et al makes it clear that research on a general, internationally accepted, unequivocal definition of, for instance, concepts like sludge congealing is indispensable, not only to support defining standards, but also to be able to compare mutual results on a real basis. For that reason European/international meetings like the present Workshop are of the highest common interest. Geuzens reports on the aim of the Belgian study and describes four routes to arrive at a well-defined standard for the concept of congealing sludge as part of the landfilling requirements that must be met by sludge-like waste material. The ultimate route mentioned "an eye must be kept on research and regulation in neighbouring countries and in the European (international?) context" is considered to be very important. Mechanical measurements of the consistency of sludge are performed by means of the laboratory vane test and a fall cone penetrometer. The latter device can be used for all kind of wastes.

The paper of H. Grüttner deals with the subject "optimization of techniques", in this special case in the interests of gaining a deeper understanding of the process and cost-reducing possibilities regarding sludge dewatering. For that reason a combination of knowledge on sludge dewatering, PC technology and management theory has been used. By way of illustration, a case study, based on a waste water treatment plant including centrifugation is given.
The discussion about the paper revealed, that first of all clear objectives and priorities for running has to be given to the operators, and that success in automation stands or falls with the availability of appropriate sensors.

The paper of L. Spinosa comprises research on laboratory scale by lack of information and distortion on information on the subject "Effect of stabilization processes on sludge dewatering and hygienization". Dewatering of aerobic treated and anaerobic digested sludges was investigated and reported on. The research yielded good usable results. Remarkable is that stabilization does not seem to affect dewatering properties of the sludge in a high rate, i.e. by dewatering the sludge to X % ds, the total costs of stabilization and dewatering do not depend on the rate of stabilization and make no difference.

The remaining part of the papers apply to practical situations.

J. Hall describes a new approach integrating waste treatment and management. The paper comprises some economic and quality aspects of composting sewage sludge, making use of straw which that is excessively available in the UK, where straw is seen as a waste material. The most important aspects like storage and handling, odours, bulk and weight, disposal outlets etc. came up for debate. In this case there is also a question of connection with the field of Working Party 4. Also discussed was the important fact to get the correct compromise in using sewage sludge and straw together (7 % straw on a wet weight basis in the UK). The volumes of straw and sludge seem to be difficult to calculate. Reuse and apliability will be only restricted by the process temperature (disinfection). Some participants question the existence of an agricultural market for the final product.

The paper of G. Boari et al on integrated stabilization processes of sewage sludges and urban solid refuses by co-landfilling and on co-digestion of sludges and leachates is an interesting example of the trend to realize co-treatment and co-disposal for optimizing techniques and reducing costs. The integrated utilization of different processes permits a better energy recovery and environmental impact control. Investigations show that the stabilization rate of the co-disposed organic wastes significantly increases with benificial effects on leachate quality and gas production. Co-digestion of sludges and leachates have a better energy recovery. Further more the paper explained a demonstrative project, financed by the Italian Government, based on an integrated utilization of three facilities to treat refuse from the Potenza municipality (two incinerators, a sanitary landfill, and the anaerobic section of a municipal waste water treatment plant, where leachates will be digested for research).

The paper of J. van den Berg et al presents a project in preparation for the combined treatment of municipal wastes, sewage sludge and liquid manure. The concept involves separation/extraction from the solid wastes of a fraction which can be used as refused derived fuel, generating of energy by incineration of the refuse derived fuel and use of the generated energy for the treatment of sewage sludge and liquid manure.
At the planned location the combined processing of waste streams has logistical, energetic, economic and environmental advantages. The main target however is to save energy.

The paper of A. Stenrroos et al summarizes a pilot scale research into the WABIO process, which is a two–stage biological method for municipal wastes and sludges. In the process organic matter is converted into biogas which can be used as fuel. Volume and weight of remaining waste material are significantly reduced. Municipal solid waste is pretreated so that organic compost fraction and refuse derived fuel are separated. The compost fraction is biologically decomposed in the WABIO process. Refuse derived fuel has a high thermal value and can be incinerated. A volatile solids reduction of 72 % has been found in anaerobic systems, compared with 50 % in aerobic systems.

In conclusion participants must have felt confident, that the fundamental and practical research on sewage sludge necessarily must go on. Nowadays sewage sludge management means more than considering treatment by its merits alone; it may be profitable to look for efficient and effective integrated treatment and disposal methods in connection with other waste materials.

LIST OF PARTICIPANTS

BOARI, G.
Universita della Basilica
I 305100 POTENZA

CAMPBELL, H.
Waste Water Technology Centre
Environmental Protection Service
Environment Canada
P O Box 5050, 867 Lakeshore Road
CANADA BURLINGTON-ONTARIO
L7R4AE

COLIN, F.
Institut de Recherches Hydrologiques
11 bis, rue Gabriel Pèri
F-54500 VANDOEUVRE-LES-NANCY

DIRKZWAGER, A.H.
Rijkswaterstaat
Dienst Binnenwateren/RIZA
P B 17
NL-8200 AA LELYSTAD

GEUZENS, P.
Nuclear Research Centre
SCK/CEN
Boeretang 200
B-2400 MOL

GRÜTTNER, H.
Forskningscentret
Water Quality Institute
11, Agern Allè
DK-2970 HORSHOLM

HALL, J.E.
WRc Medmenham
Henley Road
Medmenham , P O Box 16
UK-MARLOW, BUCKS SL7 2HD

KUTINLAHTI, H.
Outokumpu EcoEnergy
PB 86
SF-02201 ESPOO

LANGGAARD, L.B.
Water Quality Institute
Agern Allè 11
DK-2970 HORSHOLM

LEVINEN, R.
National Board of Waters and the Environment
PB 250
SF-00101 HELSINKI

NEWMAN, P.J.
WRc Medmenham
Henley Road
Medmenham, P O Box 16
UK-MARLOW, BUCKS SL7 2HD

NIEMELÄ, A.
Plancenter Ltd.
Opastinsilta 6
SF-00520 HELSINKI

OCKIER, P.
Waterzuiveringsmaatschappij van
het Kustbekken
Sandvoordestraat 375
B-8400 OOSTENDE

PAULSRUD, B.
Aquateam
Norwegian Water Technology Center
P O Box 6326 - Etterstad
Bertrand Narvesensvei 2
N-0604 OSLO 6

PERSARI, J.
Provincial Government of Kymi
PB 301
SF-45101 KOUVOLA

PUOLANNE, Y.J.
National Board of Waters and the Environment
PB 250
SF-00101 HELSINKI

SANTORI, M.
Instituto di Ricerca sulle Acque (IRSA)
CNR
Via Reno 1
I–00198 ROME

SPINOSA, L.
Instituto di Ricerca sulle Acque (IRSA)
CNR
Via F de Blasio 5
I–70123 BARI

STENRROOS, A.
Outokumpu EcoEnergy Ltd
PB 861
SF–02201 ESPOO

VAN DEN BERG, J.J.
Grontmij NV
PO Box 203
NL–3720 AE DE BILT

VÄNNI, P.
Viatek Ltd
PB 4
SF–02101 ESPOO

WALLINMAA, S.
Plancenter Ltd
Opastinsilta 6
SF–00520 HELSINKI

INDEX OF AUTHORS